CHERNOBYL

History's Worst Nuclear Accident.
The True Story of One of the Twentieth
Century's Greatest Disasters

Adam Andrews

Text Copyright © 2019 by Adam Andrews
All rights reserved. No part of this guide may be reproduced in any form without permission in writing from the publisher except in the case of brief quotations embodied in critical articles or reviews.

Legal & Disclaimer

The information contained in this book and its contents is not designed to replace or take the place of any form of medical or professional advice; and is not meant to replace the need for independent medical, financial, legal or other professional advice or services, as may be required. The content and information in this book has been provided for educational and entertainment purposes only.

The content and information contained in this book has been compiled from sources deemed reliable, and it is accurate to the best of the Author's knowledge, information and belief. However, the Author cannot guarantee its accuracy and validity and cannot be held liable for any errors and/or omissions. Further, changes are periodically made to this book as and when needed. Where appropriate and/or necessary, you must consult a professional (including but not limited to your doctor, attorney, financial advisor or such other professional advisor) before using any of the suggested remedies, techniques, or information in this book.

Upon using the contents and information contained in this book, you agree to hold harmless the Author from and against any damages, costs, and expenses, including any legal fees potentially resulting from the application of any of the information provided by this book. This disclaimer applies to any loss, damages or injury caused by the use and application, whether directly or indirectly, of any advice or information presented, whether for breach of contract, tort, negligence, personal injury, criminal intent, or under any other cause of action.

You agree to accept all risks of using the information presented inside this book.

You agree that by continuing to read this book, where appropriate and/or necessary, you shall consult a professional (including but not limited to your doctor, attorney, or financial advisor or such other advisor as needed) before using any of the suggested remedies, techniques, or information in this book.

Table of Contents

Introduction..7

CHAPTER 1: The Historical Situation At The Time Of The Disaster...11

How A Nuclear Power Plant Is Structured.............13
Reactor Number Four..15

CHAPTER 2: The Security System Of A Nuclear Power Plant..17

CHAPTER 3: 1:23:40 am..45

The Technical Aspects Of The Accident And Why It Occurred..47
Wrong Emergency Response................................51
Evacuation Management.......................................53
The Investigations.. ...56

CHAPTER 4: Victims Of The Disaster And Health Consequences..59

Testimonials...64

CHAPTER 5: Negative Consequences....................73

For The Future.......................................73
For The Area..76

CHAPTER 6: Consequences For The World.........87

How The Area Is Supervised Today......................110
Which Countries Have Suffered Other Nuclear Accidents..113
Government Declarations...................................123
Which Countries Have Given Up Nuclear Power Plants?..138

CHAPTER 7: Film and Television Productions .. 173

Conclusion 189

Bibliography 195

Introduction

"Man is harder than iron, stronger than stone and more fragile than a rose"

— Turkish Proverb

The atomic bomb dropped on Hiroshima weighed about 4.5 tons in total. It carried about 115 pounds of uranium. When the fourth reactor exploded it contained almost 200 tons of nuclear fuel and about 1,800 tons of radioactive graphite. Add to that the radioactive building structure and the radioactive water from the cooling systems.

When the bombs at Hiroshima and Nagasaki exploded most of the damage was from the impact and firestorms. The radiation was confined to a small area. In contrast, Chernobyl produced a relatively small explosion. Workers in the plant the night of the accident lived to tell the tale.

The explosion and fire produced a radioactive cloud that went high up into the Earth's atmosphere and circled the northern hemisphere twice. Rainfall deposited radioactive isotopes all over Western and Eastern Europe. Silent and invisible, the damage done by Chernobyl can't be detected without special equipment. Contrary to all the jokes, it doesn't glow in the dark.

Right from the beginning the effects of Chernobyl were influenced by politics. The Western press repeated rumors of mass graves and thousands dead. The Soviet government denied that anything of consequence had happened and accused the

West of anti-Soviet hysteria. Medical records related to the accident were classified.

One of the victims of the explosion was the nuclear power industry. Within days of the accident the American press published reassurances that U.S. reactors were safe. Experts declared that the superior designs of American nuclear plants ruled out an accident of the same magnitude. But enthusiasm for nuclear power had already been weakened by the accident at Three Mile Island where a nuclear power plant in Pennsylvania had a partial meltdown.

CHAPTER 1: The Historical Situation At The Time Of The Disaster

"Study the past if you would define the future."

— Confucio

The Chernobyl nuclear disaster's impact goes far beyond revealing incompetency in nuclear power stations and exposing the infallibility of Soviet technology as a mere mirage. It also brought to light the faults of a political system which was failing in more ways than one, showing the disastrous effects of its culture of secrecy and tendency to value individual rights below those of the group. This resulted in the population losing confidence in the regime and saw Gorbachev's programme of transparency enter into a violent collision with nuclear catastrophe. This at once fuelled policies of transparency and showed the difficulties of actual implementation; while information about Chernobyl became gradually more accessible, it was nonetheless under constant surveillance.

The catastrophe, which became a topic of national dispute, helped shape Ukraine's secession from the USSR and the fall of the Soviet Union which followed. Separatist movements among the different nations of the USSR are one of the key factors leading to its dissolution. However, Ukraine was second only to Russia in terms of how the constituent nations ranked in importance, making its potential secession a major issue. Protests began in early 1987 as the full extent of the catastrophe was being revealed and continued throughout the following years. Overall, the disaster served as a platform for opposition to the regime and as a means to awaken the Ukrainian national consciousness.

How A Nuclear Power Plant Is Structured

The Chernobyl accident amplified those concerns and plans for new reactors met with fierce opposition. Demand for new nuclear power plants ground to a halt.

The situation seems to be changing. Concern over dwindling supplies of oil and natural gas and the effects of greenhouse gasses have revived interest in nuclear power. Proponents argue that it is an environmentally sound alternative to producing electricity with fossil fuels.

His argument is reasonable. Every form of power production has health and environmental costs. The relative costs and benefits need to be weighed before any conclusions can be reached. That's why it's important to understand the causes and effects of the Chernobyl disaster.

We all live with a certain amount of risk. We get into our cars everyday knowing full well that car accidents frequently kill people but it's a risk we are willing to take. We don't analyze statistics on traffic fatalities but we have a feel for the risk involved. We see accidents on the road. We read about them in newspapers and watch local TV reports on particularly nasty accidents. It's impossible to deny that driving a car is risky but we accept that as a reasonable risk given the benefits.

In order to make a decision about the cost of nuclear power we need to look at the effects of accidents. If we don't understand what we're risking, it's impossible to make any sort of informed decision.

If you've never been to the areas contaminated by Chernobyl's radiation it would seem like a reasonable conclusion but it's impossible to believe after a visit to the area. It is impossible to know the full extent of the Chernobyl catastrophe through experience because we don't live that long. This is an event that will play out over thousands of years. Radioactive isotopes decay and form other isotopes. Some of them are less dangerous than the original. Some of them are more dangerous. As the various isotopes decay the picture changes. We know the rate of decay and the isotopes in the decay chain. What we don't know is how this will affect the soil, plants and people living in the area.

Reactor Number Four

In general, nuclear reactors produce energy through the process of nuclear fission which causes a controlled and sustained nuclear reaction. A subatomic particle called a neutron is fired towards the nucleus of an atom which splits in two and releases other neutrons and a large amount of heat, and therefore energy. The neutrons emitted in the first reaction cause nuclear fission to occur in other nuclei, thus creating a chain reaction. This chain reaction can be stabilized by permanently capturing some of the free neutrons. Nuclear reactors are joined up to generators which transform the heat emitted during the reaction into electricity.

Chernobyl's reactor no. 4 consisted of a concrete cavity with a graphite core which slowed down neutrons and thus facilitated nuclear fission. In this core, there were almost 1700 uranium rods and 200 control rods to absorb excess neutrons. These control rods could be raised or lowered to slow down or accelerate the nuclear chain reaction.

CHAPTER 2: The Security System Of A Nuclear Power Plant

"Power has only one duty - to secure the social welfare of the People"
— Benjamin Disraeli

Nuclear power plants with this type of cooling system have a positive void coefficient. Which means that when the reactor's power level increases and more steam is produced, the additional voids – steam bubbles – enable more of the nuclear particles to interact with Uranium atoms to release more heat and nuclear particles.

Reactors with a negative void coefficient are much more sensible. A negative void coefficient means that when the reactor's power level increases, the additional voids cause fewer of the nuclear particles to interact with Uranium atoms, essentially applying a "brake" to the nuclear engine.

Variations on this same type of test had been conducted before. A group of engineers and operators had carefully planned out the experiment. They had contingency plans for all kinds of scenarios. They wrote down exactly how the test should be conducted and then they handed the plans off to an engineer, Anatoli Dyatlov, who hadn't participated in the planning meetings.

Because Dyatlov hadn't been in those planning meetings, he wasn't aware of all the possible pitfalls. This, according to Dave Lochbaum, an expert in nuclear safety with the Union of Concerned Scientists, is still how tests are conducted in nuclear power plants in the United States. When I interviewed Lochbaum he said, "I've been part of that team that develops nice neat tests and then handed it over to someone who's blind to the whole thing and I've been on the receiving end of that. We can talk about the differences between Soviet design and ours but it was the human element that was such a big factor in that accident and our reactors are run by humans."

Dyatlov looked over the plans and did a very human thing; he decided to change them a bit. He shut off the automatic safety systems. In retrospect it didn't seem like a good idea. It caused two problems to combine. The experiment involved cutting power to the water circulation equipment and the reactor design causes it to accelerate rapidly if the cooling water is lost.

Normally the emergency systems would flood the reactor with cooling water if it became too hot or unstable. But Dyatlov worried that a sudden flow of cool water into the hot reactor would shock it and cause damage. The valves on the water lines feeding into the reactor were not only shut off, they were put under lock and key to prevent anyone from manually opening them.

He didn't think that shutting off those emergency systems would be disastrous because he knew that if all else failed, they could immediately shut down the nuclear reaction by lowering all the boron control rods. To understand his confidence in the control rods you have to look at how, basically, the reactor worked.

The general design is fairly straightforward. The core consisted of a 23-foot high block of graphite with long holes in it. Graphite is very similar to coal. They both are mainly carbon atoms.

Uranium fuel rods slid down into the core. Boron control rods could be pulled out or pushed in. Uranium emits high-energy particles. As the nuclear particles fly out of the fuel rods they release energy in the form of heat. They act the way a cue ball acts on a pool table. When a moving particle hits another Uranium atom it forces another particle out. Graphite slows down the nuclear particles, making them more likely to hit Uranium atoms to release energy and additional particles. The more particles there are flying around, the more energy is released and the hotter the reactor gets.

The heat from all these high-energy encounters is used to boil water and make steam. The steam is used to turn turbines that produce electricity. The metaphor of the teakettle really isn't all that far off.

Boron absorbs the nuclear particles emitted by the uranium fuel rods. The Chernobyl plant had a total of 211 control rods. According to the design specifications a minimum of 30 control rods should be somewhere in the reactor core at all times.

If all of the control rods were lowered into the core, the reactor would shut down completely. The boron would absorb so many particles that it would stop the nuclear reaction. It seemed like a failsafe way to shut everything down. However, a flaw in the design of the control rods made the reactor behave in a way Dyatlov didn't expect.

The control rods weren't solid boron. Part of the rod was hollow and part was made of graphite. The boron section, the component that absorbed the nuclear particles and slowed the reaction, was 16.4 feet but the reactor core was 23 feet. There were hollow segments above and below those absorbing segments.

The hollow space didn't matter once the rod was in place but it caused a problem when the control rod was being lowered into the core's channel. The channels are cylindrical tubes that guide the control rods. As the rod's hollow portion entered the core, it displaced some of the water that was slowing the nuclear reaction. This created a small power surge as the particles flew unchecked through the void created by the empty part of the control rod. The surge ended when the absorbing part, the boron part, of the control rod pushed into the core. Under normal circumstances the surge was negligible.

Dyatlov may or may not have known about control rod design flaw. The KGB knew about it but it was classified information. Still, people tend to talk to each other even when it's not allowed. Experienced operators probably knew about the power surges but they weren't on duty that night. The practice was to transfer the best workers to the newest plant. The most skilled employees had been transferred to the fifth reactor in anticipation of its startup.

Operators with much less experience were on duty the night of the accident. The main man at the control that night, Leonid Toptunov, was only 26 years old. He didn't have enough experience to know when to disobey Dyatlov, the engineer running the experiment. Like everyone else who worked there he had been told that nothing he did could cause the reactor to explode. No one knew how much was at stake because no one talked about accidents at nuclear power plants.

In his account of Chernobyl Grigori Medvedev, former deputy chief engineer at the plant writes, "For thirty-five years people did not notify each other about accidents at nuclear power stations and nobody applied the experience of such accidents to their work. It was as if no accidents had taken place at all: everything was safe and reliable."

The experiment had been scheduled for 2pm the previous day but it was delayed because a load dispatcher in Kiev said he needed the power. So the experiment took place in the wee hours of the morning, with a groggy, young operator, Toptunov. Everyone was anxious to get the test over with because they had a long weekend coming up and they wanted to go home.

There was one more complication. The reactor had to be brought down to a fraction of its normal power. A computerized system monitored the heat and radiation levels inside the reactor. It raised and lowered the control rods to keep the reactor stable and the power at normal operating levels. To lower the power levels, they had to shut off the automatic system.

While switching over, Toptunov was surprised by a sudden power drop from 1,500 megawatts to 30. When this happens, decay products like xenon and iodine form inside the reactor and make it slow. Like boron, xenon and iodine absorb nuclear particles. At this point, all kinds of experts agree, the experiment should have been stopped. But it would be a year before they'd be able to schedule it again. They decided to keep on going.

Toptunov pulled out one control rod after another in an attempt to get the power back up. He needed to remove the boron in the control rods from the reactor core to compensate for the xenon and iodine being formed. When he got to the point where he'd have less than the recommended 30 rods, he balked and told Dyatlov that he wouldn't take out any more. The head of the shift, Alexandar Akimov, had been watching the experiment and he agreed with Toptunov. Dyatlov called Akimov "a lying idiot." He berated Toptunov and threatened to fire him if the didn't obey orders.

Later, as he was dying in the hospital Toptunov said that he kept thinking he got to keep the job and not get dethroned. He really liked his job so Toptunov took out 205 of 211 control rods.

With only 6 control rods left in the reactor, it became hotter and hotter with more speed than anyone expected. The cooling water was boiling off, turning to steam and increasing the pressure inside the reactor. Toptunov shouted that the reactor power had reached an emergency stage. Akimov shouted back," Shut down the reactor!" He dashed to the control desk and hit the button to lower all the control rods.

The power surge caused by the hollow space in the control rods wasn't negligible when that many were lowered all at once. The graphite tips and hollow segments displaced a big part of the remaining cooling water in the core. There was nothing to absorb the nuclear particles and slow the reaction. Any remaining water became superheated. With nothing but steam and empty air to absorb the neutrons, the nuclear reaction accelerated. It was like hitting the brakes in a car and having it accelerate faster instead of slowing down. The power level of the reactor core skyrocketed from less than 10 percent to well above 100 percent in fractions of a second.

Akimov looked at the instruments and realized that the descending control rods had stopped. He disconnected the motors to let them fall from their own weight. They didn't budge. The core had become so hot it buckled. The channels for the control rods warped and the control rods got stuck at about 6.5 feet. There was nothing to slow down the chain reaction.

What exactly happened inside the reactor is still not fully understood. The official story is that superheated water reacted with the zirconium cladding on the fuel rods and produced hydrogen, which collected inside the reactor and then exploded. But not everyone agrees with this scenario.

Konstantin Checherov was one of the first people to go inside the reactor after the explosion and he has spent a great deal of time there. In fact it's one of his favorite places. He says, "When I enter the fourth reactor no one can bother me, no one can get a hold of me. There are no people around checking the radiation dose that I get there. I'm in another world, a world of freedom." He's led many expeditions inside the ruined power plant and seems to be immune to radiation. He doesn't think the hydrogen scenario works for a number of reasons.

He doesn't believe that there was enough time for a sufficient quantity of hydrogen to collect. There wasn't a space big enough inside the reactor to hold the amount of hydrogen needed for such an explosion. Further, the presence of steam from the cooling reservoirs would have dampened any hydrogen and oxygen reaction.

Checherov has his own theory about what happened after those control rods got stuck. He's studied the way the flooring and the fuel melted and what he saw convinced him that the inside of the reactor became as hot as the surface of the sun. The water vapor turned to plasma, a state of matter where all electrons have been stripped from the atoms. The pressure in the active zone, the part with the fuel rods, control rods and graphite, reached 2,000 atmospheres. This caused the first explosion as the active zone blasted up and out the reactor chamber.

A vacuum of less than one atmosphere followed the high-pressure wave. The vacuum wave engulfed the graphite core. Pressure from naturally occurring air bubbles inside the graphite pulverized the core and created a coating of graphite dust.

It's almost impossible for anyone without a doctorate in physics to determine who has the correct scenario. But it's certainly interesting that there isn't a definitive explanation of how the explosion occurred.

What was left of the graphite core burned for 10 days. It spewed out a huge cloud of radioactive particles that went up so high that commercial aircraft at 20,000 feet had to go around it. The particles caught on the wind currents in the upper atmosphere and dispersed all over the Northern Hemisphere.

Toptunov died on May 14th from severe radiation disease. He was buried in a zinc lined coffin to protect the surrounding soil from his radioactive body. It was a slow and painful death.

After all patients suffering from acute radiation sickness died, the hospital had to be decontaminated. They pulled up the flooring, tore out all the woodwork and scraped the walls.

The firemen and workers at the plant bore the brunt of the initial blast of radiation but everyone living in the Northern Hemisphere has been touched by Chernobyl. It's hard to believe that one safety test could go so wrong but it wasn't the only sacrifice made to get the plant running.

Parashin and the other administrators knew that there were construction defects but bonuses and promotions were linked to hitting the project deadlines. He says, "There was the priority of energy production over safety. This was true in the entire Soviet Union, quantity over quality. Quantity at any cost."

Designers of nuclear power plants in the United States and Western Europe have stated that even if a Chernobyl sized explosion happened at one of their sites, the containment vessel that surrounds the reactor core would hold in the radioactive material. If Checherov is correct, then it's difficult to see how any containment vessel could have withstood the heat and pressure.

The effectiveness of a containment vessel is just one technical problem in a long list of problems that lined up in just the right way to cause the disaster. Everyone agrees that the accident should never have happened. The combination of events leading to the accident was completely unanticipated. If the control rods didn't have those graphite tips and hollow sections, the emergency shutdown would have stopped it. If the operators hadn't taken out so many control rods, the reaction wouldn't have gone out of control. If they hadn't lowered all the control rods at once or if the safety systems hadn't been disconnected or if the cooling system had a negative void coefficient or if the operators weren't intimidated or... The list goes on but take out just one or two factors and massive devastation could have been avoided.

It wasn't that the event was completely unprecedented. There had been accidents and radiation releases at the Chernobyl station prior to the big one in April of 1986. Ukraine's SBU, the successor to the KGB, declassified 121 documents about the Chernobyl plant in 2003. The documents describe 29 accidents between 1977 and 1981. One refers to an inspection just weeks before the disaster where engineers said that the plant was too dangerous and should be shut down. But in 1986 all of this information was classified and the people who worked at the plants were sworn to secrecy.

The public front for the Soviet nuclear program was that everything was in the hands of the most competent professionals in the world. Underneath, the workers knew there were problems but they couldn't speak publicly about the problems. They were afraid that they would lose their jobs.

It was comforting for Parashin to believe that an accident like Three Mile Island could happen only in a capitalist country. Patriotism and years of pro-Soviet propaganda blinded him to the problems in his own nuclear industry. The political and social dynamics in the United States are different but it seems to be producing the same sort of blindness where it is widely believed that accidents like

Chernobyl could only happen in repressive authoritarian countries where secrecy and corruption are rampant.

Dr. Kymn Harvin had never given nuclear power much thought until she started working at the Salem/Hope Creek nuclear power plant in New Jersey. Once she was there she says she quickly adopted the mindset that everything was safe and anyone who believed otherwise was just misinformed.

As an Organizational Development expert her job was to improve the corporate culture. A new president had just taken over and she says he wanted to shift the culture from "kick down and kiss up" to a "much more of a healthy participative management style." For the first four and half years she worked long hours but enjoyed her work because she could see that it made a difference to the people working at the plant.

Every two years the plant was evaluated by INPO, the Institute of Nuclear Power Operations. The nuclear industry created INPO measure U.S. excellence. Although details of the 2002 INPO report are not available to the public, Dr. Harvin says it criticized the plant for not being as profitable as it should have been.

The chairman of the board made it clear to the president and all the executives that the bottom line had to improve or heads would roll. No one except the Chairman knew that PSEG, the company that owned the plant, was planning a merger with Exelon, one of the most profitable electricity and gas companies in the United States. The chairman stood to make quite a bit of money.

The working environment at the plant deteriorated and although there were serious safety concerns, no one wanted to risk losing their jobs. The workers at the plant begged after the Three Mile Island accident to reactors against established standards of Harvin to make sure that upper management heard their concerns about problems at the plant. At first, she was as reluctant as everyone else. But she knew she couldn't continue to stay silent when a man who was usually very macho took her aside and said with tears in his eyes, "I've been here a long, long time. I've never seen it this bad. You've got to help us."

She went to the President and the Chief Nuclear Officer. After all, the company's motto was, "Safety First." She assumed that even if the lower level managers dismissed the complaints, upper management would force them to deal with the problems. To her surprise, they weren't interested in the worker's concerns. They were already aware that there were serious hazards but another shut down would cut down earnings. A director at the plant even told her that if the NRC, Nuclear Regulatory Commission, knew what they were doing, they would, "take the keys away."

After she lost her job she was invited to evaluate the Operations departments at three top-rated nuclear plants. The contrast between the conditions there and at Salem/Hope Creek made her realize how dangerous the situation had become.

After much soul searching she decided to go directly to the nuclear police, the NRC The NRC's mission is to protect the American public from unsafe conditions at nuclear power plants. If the management at the plant wouldn't listen, the NRC would step in. She also initiated a whistleblower lawsuit.

The NRC launched an investigation and employees at the plant verified Harvin's claims. The NRC wrote a letter to the president asking him to improve the safety culture. There were no other sanctions.

After they realized they couldn't ignore her, PSEG offered her a sizeable settlement. Dr. Harvin had lost her job and no one in the industry would hire a whistleblower. But she could see that they weren't addressing the safety issues so she turned down the settlement.

During the discovery phase of her trial she found out that dozens of other workers at the plant had come forward with serious problems. Everyone else had taken settlements that ranged from $40,000 to $750,000 and agreed to keep quiet.

While this came as a surprise to Dr. Harvin, it wasn't news to Dave Lochbaum, an expert in nuclear safety with the Union of Concerned Scientists. Lochbaum worked with Harvin on her lawsuit and he's worked with many other whistleblowers. He says that it's not uncommon for companies to pay off troublemakers, as long as gag orders are written into the settlements.

Lochbaum has a degree in nuclear engineering and started his career in the nuclear industry as an operating engineer. He believed in nuclear power then and he still believes that, if the plants are operated safely, it is a useful technology.

He joined the Union of Concerned Scientists after he went through his own whistleblower ordeal. In 1992 he tried to get a problem with the cooling system at the Susquehanna nuclear power plant addressed. There was a serious design flaw in the cooling system. Lochbaum realized that, under the right conditions, it could lead to a meltdown.

The plant owner wouldn't address the issue so he went to the NRC and they didn't act on the information. Lochbaum and his colleague, Donald Prevatte, went to congress and the media. After a cover story in Time magazine brought pressure on the government to act, the design was changed.

It's easy to look at the settlement numbers and think that whistleblowers have come out ahead. But when the money is weighed against the loss of a career, the years of waiting in legal limbo and the intimidation most people regret the decision. That was the conclusion of C. Fred Alford in his book, "Whistleblowers: Broken Lives and Organizational Power." He found that the majority of whistleblowers lose their jobs; many lose their families and go bankrupt. They also suffer from depression and alcoholism. Harvin says she's lost many friends, suffered a deep depression and was left, "one step short of shattered" by the experience.

Lochbaum has seen the harsh retaliation in the nuclear industry and he says that he can't fault anyone for taking cash settlement to keep quiet. Harvin is considering a settlement. She can't get a job in the nuclear industry and even after years of fighting the lawsuit, she says that conditions at the plant haven't improved much. It pains her to think about how much of her first amendment rights she'll give up.

It isn't true that harassment and intimidation are the industry norm. Lochbaum and Harvin both can cite many well-run companies where employees are encouraged and even rewarded for speaking up. In fact, Lochbaum says he hasn't seen a single incident where intimidation has worked in the long run. Sooner or later the problems become too severe to keep covering them up. The problem is that all it takes is one Chernobyl and thousands of lives are put at risk.

Lochbaum isn't optimistic about the future of nuclear energy in the United States. He sees the NRC as ineffective. Instead of acting as watchdogs, they partner with the industry. At the same time, that the Bush administration is providing billions of dollars for building new plants, there are no plans to put more money into safety enforcement.

The risk of a serious accident will increase in the next few years as the existing nuclear plants age. Rust and cracks take their toll on the equipment. At the same time, new plants will come on line. Lochbaum says that all the major accidents, so far, have happened in the first year or two of operation.

The combination of aging plants, new construction and lax oversight creates ideal conditions for a Chernobyl right here in the United States.

Still, Harvin and Lochbaum insist that they aren't antinuclear, they just want the industry to operate under safe conditions.

In Ukraine, Sergei Parashin is also still enthusiastic about nuclear power. For many years after the accident he even retained a career in the nuclear industry.

In 1994 he was made director of the Chernobyl station where thousands of people worked. The fourth reactor was ruined but the other reactors operated, on and off, until 2000. Also because there is still fuel there so the station won't be completely shut down, according to Checherov, for about a hundred years.

Parashin inherited a dispirited work force when he took over at Chernobyl. After carefully assessing the situation, he decided that people weren't radiation. Bad attitudes were person's psychological state affects their immune system, "People with positive, optimistic attitudes - they tend to live longer even though they've been working at the station and have been exposed to radiation. People who are a little bit negative about their lives don't live long here."

To address the root of the problem he redesigned their uniforms to look less like prison uniforms. He got a better grade of utensils and nice napkins in the dining hall. They planted flowers and put up fountains.

CHAPTER 3: 1:23:40 am

"My eyes have become dark because of my pain, and all my body is wasted to a shade"

— Job 17:7

The Chernobyl nuclear disaster's impact goes far beyond revealing incompetency in nuclear power stations and exposing the infallibility of Soviet technology as a mere mirage. It also brought to light the faults of a political system which was failing in more ways than one, showing the disastrous effects of its culture of secrecy and tendency to value individual rights below those of the group. This resulted in the population losing confidence in the regime and saw Gorbachev's programme of transparency enter into a violent collision with nuclear catastrophe. This at once fuelled policies of transparency and showed the difficulties of actual implementation; while information about Chernobyl became gradually more accessible, it was nonetheless under constant surveillance.

The catastrophe, which became a topic of national dispute, helped shape Ukraine's secession from the USSR and the fall of the Soviet Union which followed. Separatist movements among the different nations of the USSR are one of the key factors leading to its dissolution. However, Ukraine was second only to Russia in terms of how the constituent nations ranked in importance, making its potential secession a major issue. Protests began in early 1987 as the full extent of the catastrophe was being revealed and continued throughout the following years. Overall, the disaster served as a platform for opposition to the regime and as a means to awaken the Ukrainian national consciousness.

The work of the operative group became more concentrated on the construction of the cover that was to encase the destroyed Unit 4. It was called "Shelter of the Fourth Unit," but later—by the light hand of a writer—it got a new name: Sarcophagus.

The Technical Aspects Of The Accident And Why It Occurred

We have spoken already that on the sixth of May, ten days after the accident, the release of radioactivity out of the destroyed Unit, which threatened to cause serious disasters, decreased suddenly by hundreds of times. At the time, the reason for this decrease was considered to be the effect of all the materials that had been thrown from the helicopters. Now we know that those materials hadn't played their proper role. The explanation is different now. At that time, the fuel, having melted the lower protective plate of the reactor, dissolved in other melted materials and formed radioactive lava, never seen in nature before. Lava spread in the lower stores of the block and started to cool. The radioactive release was practically stopped.

China Syndrome—the burning of concrete plates and the gradual falling of the fuel— worked only for that lower plate and, to some extent, for the floor of the room situated right under the reactor.

Again, we only know this now. Then, in May of 1986, only one thing was absolutely clear: the situation was stabilizing somehow.

It was necessary to cover the open radioactive wound—the reactor's ruins—as soon as possible, to isolate it from the environment. Without this, a strong wind could spread dust out of it. Rain water could get into it, absorb radioactivity, and contaminate ground waters. And last, penetrating radioactivity threatened everybody working at the Station.

About eighteen various projects of Shelter were presented. To avoid the details, the projects can be divided into two groups. The first group consists of big constructions, covering with a hermetic cupola or hangar the whole building of Unit 4. The remaining buildings with radioactive materials would be kept inside Shelter. Thus, all destroyed buildings would be stored in the huge Shelter. The second group consists of projects that suggested using the remaining walls of Unit 4 as supports for new designs. That would greatly reduce the size of Shelter.

The second group was chosen. It ensured high benefits in time and cost of building.

I think the choice in that trying situation was absolutely right. Though, as it frequently happens, nobody thought about defects in the second group.

Nobody tried to eradicate them. Those defects were not even mentioned.

What were the defects?

To use the existing constructions that remained after the explosion and fire, partially destroyed as they were, as supports for new constructions, it was necessary to be sure of their strength and stability, which required taking measurements. That was impossible to do. High radiation fields around and inside the Unit created obstacles. The majority of measurements had to be done from a helicopter, as a guess. That is why the real extent of the strength of the new constructions was practically unknown.

To work on the building right at the block and not at a distance—as would have been done with the projects in the first group—it was possible only with the use of remote-controlled mechanisms. These mechanical devises were bought abroad quickly, unusual for our organizations. They were a number of special concrete pumps—Putzmeisters—which transported mortar through long hoses, controlled at a distance; and a number of cranes of high-carrying capacity—Demagy. These mechanisms ensured quick distance building. But ... At this building, it was impossible to use welding to form the constructions properly. And there remained numerous cracks in the original construction. Because of that, Shelter was not hermetic. Through these cracks, water and snow could get in, and radioactive dust could easily get out.

So, two main defects—that this construction wasn't hermetic and had indefinite strength—became the price for low cost and speed.

Wrong Emergency Response

The explosion happened, ironically, during a safety experiment. The reactor was being shut down for routine refueling and this gave the plant operators an opportunity to conduct a safety test. This particular test should have been conducted before the reactor was started but in the rush to get the plant operational, it was put off.

The purpose of the test was to see if the reactor could keep pumping cooling water in the event of a power black out. It seems counter-intuitive that a plant producing electricity should require power but fuel rods need to be kept cool under all circumstances. If the turbines quit producing power, back-up diesel generators would keep the pumps going. But if there were a black out, it would take a few seconds for the back-up generators to kick in. The engineer conducting the experiment wanted to see if the turbines would spin long enough to pump sufficient water through the reactor to cool it until the back-up engines could take over.

Those few seconds were crucial because when the reactor lost cooling liquid it didn't shut down. It produced more power. The increased power turned any remaining cooling water into steam inside the reactor core. With less water to absorb nuclear particles, the reaction accelerated. In this type of reactor, increased steam generation leads to a further increase in power. Once this cycle starts, the heat and energy output of the reactor can increase rapidly and make the whole situation difficult to control.

Evacuation Management

Another problem was living conditions. Before the accident, the workers lived in the town of Pripyat. The brand new city was a showcase community until it was evacuated and abandoned after the accident. Another city, Slavutich, was built for the workers. In the rush to build the new city, it was cited on radioactive land.

getting sick as a result of the real culprit because a

Many people felt that living in Slavutich was dangerous despite the fact that every precaution had been taken to protect the workers. Areas with high radiation levels were roped off. The city builders cut down a number of trees that were deemed to too hot to leave standing but they didn't want to sacrifice all of them. Trees with lower radiation levels were gently scraped and left standing.

Still, Parashin heard rumors that people living in Slavutich were worried about their children's health. Relatives from other parts of the Soviet Union would call them and say things like "You really should get your kids out of there."

Parashin tackled this problem head-on. He planted flowerbeds in the city. He held concerts with pop stars like Latoya Jackson and the French performer Patricia Casse. He organized children's festivals because, after all, it wasn't the radiation but fear of the radiation that caused the sickness.

He was dismissed from his position in 1998 because he fought against the permanent shutdown of the Chernobyl station and he openly criticized Ukraine's newly formed national energy company, Enerhoatom.

Ukraine currently operates four nuclear power plants that supply nearly half of its energy needs. Ukraine's government is seeking funding for building new nuclear reactors and is planning on importing nuclear waste from other countries. Parashin believes that Chernobyl provided a big benefit to the rest of the world because it forced better oversight and safety. In the next breath he says that this benefit hasn't really reached Ukraine. There is still a lax attitude about safety.

Ukraine doesn't have the resources to deal with another Chernobyl size disaster. The Soviet Union was able to bring together massive amounts of resources; experts from other republics with nuclear power plants, equipment and over 600,000 workers. A study done in 1991 by a Soviet nuclear industry economist estimated that the cost of the Chernobyl accident was several times greater than the net economic value of the entire Soviet nuclear industry.

Parashin still sees nuclear power as a cost effective means of producing power and he welcomed the new reactors that were put on line in 2004. He says that the new plants don't have any of the design problems that plagued Chernobyl. Now they have it right and it couldn't possibly happen again.

The Investigations

It seems old habits die hard; the local authorities released almost no information whatsoever in the immediate aftermath of the explosion. The accident was only announced two days after the explosion in a brief press release, then an equally brief mention of the incident was made on the evening news. The official statements released in the days which followed maintained a reassuring tone, repeating that the situation was continually improving. Fearing the stir it might cause in the international media, the USSR refused American help on 29 April under the pretence that all problems had been solved. Mayday celebrations took place as normal in Kiev, despite the high levels of radioactivity there at the time. Finally, Gorbachev made a televised announcement on 14 May 1986 in which the accident was officially recognized, though still played-down.

The authorities' reaction is in part due to their usual tendency to cover up problems and in part due to the difficulty of facing up to such an unexpected disaster. Evacuations began too late and no actions were taken to protect the population from the initial outpouring of radiation, despite the fact that solutions did exist. The team of liquidators also payed the price among this policy of non-recognition; the radiation doses they were received were underplayed and medical directives banning any link between illnesses they developed and their exposure were issued.

Following on from this, the accident was presented in a dramatic light to emphasize Soviet courage. The liquidation process was framed as a war against an invisible enemy; the liquidators were heroes and the authorities in charge of the process were praised for their response. Despite the danger posed by radiation, a Soviet flag was put into reactor no. 4's chimney at the end of the liquidation effort with a banner attached which read "The Soviet people are stronger than the atom".

CHAPTER 4: Victims Of The Disaster And Health Consequences

"Always the innocent are the first victims, so it has been for ages past, so it is now"
— J.K. Rowling

One of the few problems officially associated with the Chernobyl disaster is the increase in thyroid cancer in Ukraine, Belarus and Russia. A World Health Organization researcher who attended a

Chernobyl conference in 1991 reported that the official speaker for the International Atomic Energy Agency (IAEA) mentioned that this was a "good cancer". Anna Kaslova might disagree with him.

No one believes the official Soviet count of 31 deaths directly attributed to the accident but no one agrees what the number should be, either. Twenty-eight of the thirty one were firefighters who died within months. One hundred thirty four developed acute radiation sickness. A United Nations report estimates 4,000 excess cancer deaths; other estimates range much higher.

There is a big, imposing old building at number 26 Novaya Basmanaya Street northeast of the Kremlin in Moscow. Clinic number six specialized in radiation and nuclear issues, and from the start it was nearly overcome with patients, because while political Moscow showed itself ham-handed, the medical establishment proved adroit. By nightfall on the first day, only sixteen-odd hours after the blast, they'd flown 129 patients into clinic number six. 170 more followed the next day.

Specialized first responders flew to Chernobyl in half a day. Medical units gathered in Moscow scant hours after the accident, before dawn, and arrived in Kiev at 11:00 a.m. They examined over a thousand people.

Ultimately 5000 doctors and nurses descended on Polissia, especially therapeutic radiologists and hematologists. In Moscow some 300 more provided care. Forty-year studies are ongoing to compare cancer rates of evacuees to control groups. It's the largest study since Hiroshima.

The reactor 4-night crew supervisor, thirty-three-year-old Aleksandr Akimov, died from radiation poisoning in clinic number six after two weeks. To the end he remained defiant and in denial. He thought he'd done everything right.

Vyacheslav Stepanovych Brazhnik, senior turbine machinist, stayed in the turbine hall when the building exploded, trying to stabilize things. He too received a fatal dose. According to Medvedev: "Brazhnik lay there all swollen, dark brown. He strained to say that his whole body pained terribly and that he was weak....

"He was in a very bad way, and I did not want to ask him any more questions. He was constantly asking for something to drink. I would give him Borzhomi mineral water.

"The pain, pain, everywhere.... The pain could be so terrible...."

They awarded Brazhnik the order "for courage."

Viktor Vasilyevich Proskuryakov was a trainee and recipient of the same posthumous award. When Akimov began to realize the scale of the disaster that night, he implored Proskuryakov to go into the reactor building and fix it by hand. Akimov sent him to his death, with 100% radiation burns.

Grigoriy Medvedev:

"I went to see Proskuryakov 2 days before he died. He was lying on an inclined bed. His mouth was horribly swollen. No skin on his face. Bare. Bandages on his chest. Heat lamps above him. He asked constantly for something to drink. I had mango juice with me. I asked if he wanted some juice. He said that he did, very much. He was tired, he said, of the mineral water. A bottle of Borzhomi stood on his bedside table. I helped him to drink the juice from a glass...."

"The SIUR (reactor operator) Lenya Toptunov had his father on watch by the bedside. He had already given his bone marrow for his son's transplant. But it did not help. He stayed at his son's bedside day and night, he would turn him over. He was burned to blackness all over. Only his back was still light colored.

Testimonials

Vladimir Kovzelev still sees the village in his dreams but in the real world it no longer exists. Only the cemetery and war memorial remain. About five hundred people lived there before the accident. His family had lived there for over 400 years. Now weeds push up through the broken pavement and patches of black-eyed Susans and bachelor buttons mark where houses with carefully tended gardens used to stand. As the years slip by the forest encroaches and someday even that evidence will be erased.

Kovzelev didn't know about the Chernobyl accident when he marched in the May Day parade in 1986. He lived in Gomel, the biggest city near the village, and after the parade he went to his parent's house in the village where he grew up. It was hot that spring, over 80 degrees in April and May. People went out sunbathing. They had family picnics outside. Vladimir went fishing.

The first inkling that something wasn't right came to him while he waited to get a bus back to Gomel. Military transport vehicles filled with civilians blocked the roads. At work on Monday there was talk of an accident, some sort of accident but no one knew where or what kind. There was no official word, just rumors. Vladimir knew that something big had happened when he saw the evacuees pouring into Gomel with nothing but the clothes on the backs.

The next weekend he went back to his parent's house. Everything looked the same except for the soldiers from the chemical protection regiment. They walked up and down the dirt roads with dosimeters taking readings but they said it was just routine measurements. Nothing to worry about.

Then the media broadcast warnings about the radiation. People could protect themselves by keeping their windows and doors shut. Keep out the dust, the radio announced, and you'd be safe from the radiation.

***

Svetlana Polganowskaya lives in Cherchersk, Belarus. It's about as far away from the reactor as Vladimir's village and it's just as contaminated. Right from the beginning Svetlana knew that something was wrong. The day after the accident her children were playing outside when the rain started.

The wind blew sand everywhere and when Svetlana washed the children she noticed that there was a strange gray dust in the bathwater. When she went outside the puddles had a yellow/green tint. A few days later she went to the equivalent of city hall and tried to find out about the weird rain. The local authorities, under the command of Mr. Zaretsky, told her she was too clever for her own good and she should go home and keep her mouth shut. That was the beginning of her problems with Zaretsky.

In early June a very curious thing happened. The government announced that all the school children would be going on holiday together. This had never happened before. Rumors about an accident were all over but the authorities denied that the sudden vacation trip had anything to do with it. Nervous parents obediently gathered their children in the central square.

But Svetlana didn't like the idea at all and while everyone waited for the buses, her father talked about the evacuations during World War II. He wasn't the only one who remembered how families were separated and children were lost. By the time the busses arrived the parents mutinied. They held on to their children and screamed at the authorities that they wouldn't let their children go because they were afraid they wouldn't find them afterwards. The outing was cancelled and a few days later mothers were issued vouchers to go with their children to the Gorky region. They spent three months there but then returned.

Despite official denials that anything was wrong with the food or water, Svetlana was more inclined to believe the rumors. She stocked up on food whenever she went to Minsk or Gomel and avoided the local produce.

Her fears about the radiation got a boost in 1988. She was working as a shooting instructor and a group of men from the Rostov region came to her class. They told her that they were surprised to see so many children and young people around. They'd been told that only old people lived there because it was dangerous for everyone else.

As villages around them were emptied out, Svetlana grew more suspicious. Then she got some very strange information from her brother-in-law. He was a photographer in the Kremlin. He said that he'd seen papers saying that their village had already been evacuated.

In 1990 she organized a trip to Moscow with people from the area who were concerned about the radiation. If the local authorities wouldn't give them any information, she'd seek it on a federal level. When Zaretsky and his deputies found out about the trip they called Svetlana. They told her that the area was highly contaminated and if she cancelled the trip to Moscow, they promised to bring in clean food. She declined the offer.

A short while later a friend in the militia phoned her and suggested that she should move the departure time up a few hours because the militia planned to block the roads. She took the advice and they made it to Moscow. She did some investigating and found documentation that confirmed that the area was highly contaminated. It also said that housing had been provided for the evacuees and money for food subsidies were going to Zaretsky.

They never did find out what Zaretsky did with the money but she managed to get the addresses of the houses that Zaretsky and his cronies had obtained illegally. She made her TV debut on the Soviet Peace Fund telethon where she pleaded for help for the people in her area. She was contacted by an international organization that gave her dosimeters so she could measure the radiation levels and a group of sailors donated 100 kilos of caviar. They'd heard that caviar would help with the radiation.

Next she went to an international conference in Kiev and spoke about the problems in Belarus. Prior to that, most of the international aid organizations had focused on the clean-up workers at the Chernobyl plant, especially those living in Ukraine. People had no idea what was going on in Belarus.

She met a group of Japanese activists and invited them to come to her area. She showed them where combines were harvesting grain in highly radioactive areas. The Japanese were horrified. They went to Zaretsky and asked if they could monitor the radiation the way it had been done in Nagasaki and Hiroshima. Zaretsky told them that he couldn't allow them to do the monitoring but if they'd give him the money he'd do it himself. They declined.

The Saturday after the Japanese left Svetlana was at home in her apartment with her children when someone buzzed her from the outside door in the building lobby. Usually she'd just open the door but this time she took a look at them from the window. There were three big men she'd never seen before. She refused to open the door. They said they were from the KGB. She didn't let them in and she didn't leave her home for two days. She was sure that she'd be arrested if she went outside.

Finally, she phoned Zaretsky and told him that if he didn't leave her alone she'd call the Israeli Embassy and tell them that she was being persecuted because she was Jewish. The KGB never showed up again.

CHAPTER 5: Negative Consequences

"As the scorpion said to the dying girl: 'Did you know that I am poisonous when you picked me up'... "

—Stephen King

For The Future

In many ways Chernobyl is unique. The number of countries affected was unparalleled. But it wasn't the biggest radiation release into the environment and it isn't the oldest. To predict the future for people living with the fallout from Chernobyl, it's instructive to take a look at the Chelyabinsk region in Russia. It is probably the most polluted area on the planet. The people there have been exposed to radiation in the environment for over 50 years.

Environmental considerations were largely ignored during the race to construct atomic weapons and power plants. The military of both the United States and the Soviet Union put production above all else. The Hanford nuclear Reservation in Richland Washington, where the plutonium for the first US bomb was made, is an environmental disaster. There are huge tanks of radioactive and chemical waste that will take decades to clean up. But there were efforts to protect the workers and keep the radiation from spreading to the surrounding environment.

At the Mayak facility, there was little or no protection for the workers, after all most of them were convicts. Radioactive waste was dumped directly into the river Techa for the first six years that the facility operated. People living downstream were told nothing. In order to keep the processing secret, workers were not allowed to take notes or write down the steps involved in creating plutonium. They had to memorize everything. They made many mistakes resulting in leakages and explosions. In 1952 a health survey of Metlino, one of the villages near the complex, found that 67% of the inhabitants had leukemia.

The Soviet government realized they had a severe problem when radiation from the Techa River was detected in the Artic. They stopped dumping into the river and evacuated people living downstream from the Mayak complex. Radioactive waste was then dumped into an enclosed body of water, Lake Karachay.

For The Area

The measures weren't sufficient to keep the area from further pollution. In 1957 an accident occurred that rivals Chernobyl in the amount of radiation released. A cooling unit at a waste facility failed. An area of about 9,200 square miles that contained over 270,000 people was contaminated.

Since the area was closed to the outside world, people were quietly evacuated and all information about the explosion was kept secret. Although unverified accounts of the accident circulated in the Western press for decades, the incident wasn't officially acknowledged by Russia until 1992, after the fall of the Soviet Union.

A Freedom of Information request filed in the late 70's found that the CIA knew about the 1957 accident but didn't release the information. The government of the United States apparently felt it was more important to protect the nuclear industry than it was to reveal embarrassing secrets about its cold war enemies. It seems that the only purpose for the secrecy was to keep the Soviet and American citizen ignorant.

But that wasn't the last disaster. The Soviets stopped dumping waste into the Techa River after radiation from the river reached the artic. They dumped high-level waste into Lake Karachay from 1951 to 1953. Low to medium level waste was dumped there for decades. A person standing on the shore of Lake Karachay, before it was covered with concrete, would get a fatal dose of radiation in just one hour.

Dr. Gulfarida Galimova moved to the area in the 1980's. She found that half the women were sterile and the average life span was 47. She couldn't understand why everyone in the area was so ill but others knew.

Dr. Mira Kosenko head of the clinic at the Institute of Biophysics, had worked with patients from the Chelyabinsk area since the 1960's. She wasn't allowed to tell her patients that they were suffering from radiation induced diseases, "They didn't know anything, and we had no right to tell them that they were irradiated. All this information was top secret. And it was a secret because of the factory where they produced weapons grade plutonium. And no one was supposed to know its location. If someone found out that in some area there were people who had been irradiated, then it would have been possible to find the factory."

Ironically, Chernobyl may have helped people living near Mayak. After Chernobyl, descriptions of radiation induced illness spread in the mass media. People began to make connections between the illnesses. The NGO "Aigul" conducted a genetic survey of people living near Mayak. They found that one out of every four children has a genetic mutation. Even if people moved away from the area, their children carry the genetic markers of radiation exposure.

Guzman Kabirov grew up in the area and he knows firsthand the health problems of the region. He became an activist and is working to help people in the region. What he's seen is that the worst effects don't show up until the third or fourth generation, "What we have now in our place they will have in Chernobyl in 20 years."

But scientists who are studying the genetic effects of the radiation are not convinced that this is true. Boris Sorochinsky has studied the genetic mutations in pine trees growing near the Chernobyl plant. He believes that it is far too early to make any pronouncement, one way or the other on the effects of Chernobyl. "With a few scientific facts and a good speaker, you can prove anything right now."

A month and a half after the explosion Sorochinsky gleefully approached the exclusion zone, the highly radioactive area right near the reactor. His specialty is radiobiology or the study of radiation on living organisms. Since he didn't have access to the contamination at Mayak his studies had been nothing but a theoretical science until the disaster hit. Now he had an outdoor laboratory with a complex ecosystem and lots and lots of radiation.

Chernobyl surprised him again and again. The first surprise was when he switched on the dosimeter. The radiation levels were a military secret. He had no idea what he was in for until he stood in a pine forest next to the reactor with the dosimeter in his hand clicking wildly. That's when he understood the scope of the disaster. There was far more radiation than he'd expected.

The next surprise was the appearance of trees near the reactor. Everything was the wrong shape, size and color. Oak trees had leaves that were either huge, ten times their normal size, or stunted, a quarter of the normal size. The pine trees weren't green. They were yellow and red and the needles were also either oversized or abnormally small. He made measurements. He took notes. All of this was new territory.

He wanted to study the effects of the radiation on the genetic structure of living organisms. NATO financed a project where he worked with scientists from the University of North Carolina to look at the genetic changes in irradiated plants.

Plants were easier to study than animals because they stay in one place. He could control radiation levels, nutrients and water supply. He decided to study pine trees because 40% of the land in the exclusion zone was covered with forests and most of the trees were pines. He returned to the exclusion zone again and again so he could match deformities in the plants with radiation levels. He expected to see the number and severity of the deformities drop when the radiation levels dropped. Chernobyl surprised him again.

During the first year or so the he measured high levels of radiation just standing in the zone. It was in the air and water. It was right at the surface. It made sense that the exposed plants would grow abnormally. After the first year the ambient radiation levels dropped but plants sprouted in the zone still didn't look normal. Sorochinsky realized that the changes were caused by the radionuclides that the plants soaked up from the soil. The changes weren't caused by outside radiation but from the radiation inside the plants themselves.

He could hold the dosimeter to the ground and it registered radiation only slightly higher than normal. He put the dosimeter up to the leaves on a tree and it clicked fast and steady.

The plants that looked deformed grew only in the most highly contaminated area around the reactor. He found that normal looking plants, ones that grew some distance from the center of the contamination were also soaking up radiation from the soil. He wondered if the internal radiation was changing the genetic structure of the plants.

It would be impossible to look for specific mutations. Tens of thousands of genes and mutations might show up anywhere. There was no guide for where to look for specific changes. All he could do was compare the overall genetic structure of plants exposed to radiation to genetically similar plants that weren't exposed.

He took a population of plants that were genetically very close and planted some of them in soil heavily contaminated with radionuclides. Then he took some of the plants and put them in clean soil.

The comparison for genetic similarity is like the tests used to establish paternity. It compares overall similarities and differences to determine genetic distance or closeness. He found a direct correlation between genetic distance and radiation levels. In other words, the more radiation the plants absorbed the more genetic plants.

He can't say exactly occurring but he can say that clearly the radiation is affecting the genetic make-up of the plants. His concern isn't just for the trees, it's also for the people who live nearby. Sorochinsky grew up in a small town in the Policissa region. His parents still live there. He used to fish in the rivers and lakes but he gave that up after the accident.

His parents, like many of the people there, gather the mushrooms and berries in the forests. Mushrooms and berries concentrate radionuclides and he's warned his parents many times that eating them is dangerous but they still do. They say that they've done it for so long that they don't intend to stop. It's a tradition that they don't want to let go of.

CHAPTER 6: Consequences For The World

"The nation that destroys its own soil, destroys itself"

—Franklin Delano Roosevelt

The official position of the International Atomic Energy Agency and the World Health Organization is that there are no long-term consequences. But Boris Sorochinsky doesn't share their optimism.

He believes that the genetic changes he sees in plants are also happening to people who are either living in contaminated areas or are eating contaminated food.

These changes will play out over generations. It's only been twenty years since the explosion. As far as Boris Sorochinsky is concerned, we are just beginning to see some of the consequences. He believes that it will take five or six generations before the full impact is realized.

He jokes that Chernobyl might become a new source of genetic diversity. There is no way of telling what sorts of changes will occur and if they will be good or bad, big or small.

The idea that radiation induced genetic changes can be passed on to offspring was also confirmed in the BEIR VII study by the National Academy of Sciences.

The areas contaminated with cesium 137 will need to be monitored for 300 years. The United States didn't exist three hundred years ago. The steam engine hadn't been invented and the Spanish Inquisition was still looking for heretics.

There are other isotopes, like americium-241, which is the decay product of other isotopes. So for example as the plutonium-241 decays, the amount of americium-241 will actually increase. Americium is more water-soluble than plutonium so it can move into the soil and then into plants and then into people. That means some areas, mostly right around the reactor will become more dangerous, rather than less over time. Americium 241 has a half-life of 432.7 years so using our radiobiology rule of thumb; it will remain dangerous for 4,327 years. Four thousand years ago, Rome didn't exist. It would be over 2,000 years before Christ made trouble for the Romans.

How dangerous the area will remain is another unknown. If the isotopes don't spread, the only affected areas will be ones that are already depopulated. The biggest danger is if it becomes mobile.

Studies done in Palo Alto California during the 1950's showed that Americium could become mobile in water. However, Sorochinsky has been testing the areas around the reactor for the presence of Americium. He found that the microorganisms in the boggy soil seem to bind the Americium and keep it from spreading.

But, he also warns that this will be true only as long as the microorganisms don't change. The presence of radiation means that there will be changes. Whether these changes will be beneficial or detrimental is another one of those questions that can't be answered because there are no precedents to go by.

Kabirov has been watching the population in the area around Mayak decline over the last few years. He says that he doesn't believe the area has any future and that in 40 years or so it will be made into a national park because all the people will be dead. Whether this is the future for the Chernobyl victims remains to be seen.

Belarus was heavily contaminated by the fallout from Chernobyl and the population of Belarus is declining. But neither the government of Belarus nor the IAEA nor the WHO attributes that to radiation. The decrease is blamed on poor economic conditions, poor health care, the break-up of the Soviet Union but not on radiation from Chernobyl.

Ivan Nikitchenko was the Deputy Minister of Agriculture for the Republic of Belarus when the Chernobyl plant exploded in 1986. He was given the task of setting up a system to ensure radiation-free food production. He measured radiation levels in the food and soil and experimented with the effects of ingested radiation on farm animals. The results of those experiments alarmed him. He found that many of the cows, sheep and pigs became sterile. The ones that could become pregnant often lost the fetuses before full term. He now sees the same trend in the human population, "Since 1993 the population of our country has been decreasing. The speed of this decrease has grown year by year. We are in a zone of biological catastrophe, demographical catastrophe as well."

According to Nikitchenko, the declining population of Belarus can be traced directly to Chernobyl. Decreasing fertility and an increase in mutations that cause spontaneous abortions are to blame, "Radiation causes different mutations. So animal species, to protect themselves against these mutations don't allow them to be born."

While many debate the long-term genetic effects, another very pressing matter is developing. The sarcophagus, the building wrapped around the fourth block is crumbling. Immediately after the accident specialists from all over the Soviet Union were brought in to contain and clean up the Chernobyl site. But no one had ever even conceived of such a problem, let alone have experience with it. So they did the best they could but decisions were made quickly and without much solid information.

It became apparent as early as 1994 that the sarcophagus was not going to last. It was difficult to know exactly what was going on inside the building. People couldn't walk inside and take a look because the radiation is too high. Even a quick look could mean exposure to a fatal dose.

It's even difficult to use robots in some areas. Robots with hard-wire controls get tangled up in debris. Wireless controls won't work in the high radiation environment.

Engineers believed that the main supports for the roof of the fourth reactor were damaged in the explosion. The wall that the fourth block shared with the third block looked as if it might fall. The 500-ton biological shield had been flipped upward and no one knew if it was in a stable position. Air and water flowed in and out of the building, causing erosion and spreading more radiation.

The fuel itself is believed to be stable because it was encased in a sort of lava flow during the accident. But no one knows what will happen to it over time, especially if it comes in contact with the water that's seeping in.

Another problem is theft. In 2002, five men were put on trial in Belarus for attempting to sell Uranium in Minsk. Details about the case are difficult to find and confirm but apparently they were caught with about 1.5 kilos of uranium that came from Chernobyl. It wasn't enough fuel or refined enough to make a nuclear bomb but it could be used as a "dirty bomb."

The EU wanted Ukraine to permanently shut down all the reactors at the Chernobyl site. Reactors one, two and three at the station are the same design as the one that exploded. Europeans living several thousand miles away knew that another accident could affect them. But Ukraine was reluctant to shut down all the reactors because they needed the power.

In 2000, the EU prevailed and Ukraine closed down all the Chernobyl reactors. As part Ukraine would get assistance in plants where construction was halted after Chernobyl. They of the shutdown plan, finishing nuclear power also agreed to allow an international consortium to build a containment facility over the damaged reactor.

The plan is to create a series of arches that will be moved over the plant. The idea is to keep the rain out and radiation in. The arch, also known as the Shelter Project, has an estimated cost of $1 billion. It's scheduled to be completed by 2011 and designed to last 100 years. Scientists working at the reactor estimate that it will take 75 years to clear out all the fuel.

Critics of the program say that the arch will do nothing but cover up the problems and make Europeans feel better without addressing the core problems. In addition, the current design will expose thousands of workers to high levels of radiation during the construction and maintenance of the facility. "It's not going to be western people who are going inside the sarcophagus to work. It's going to be Ukrainians working for minimal wages," said Vladimir Kostenko who worked for many years in the public relations department of the Chernobyl plant.

Kostenko left his position after publishing an article in a Russian newspaper that questioned the details of the project. He believes that safety and environmental problems haven't been properly addressed and that the new construction will delay and hinder a real clean up. The one point that everyone involved agrees on is that the current situation is unstable and something needs to be done.

On top of the physical structures that need to be rebuilt, there is an entire social structure that was blown apart by Chernobyl. Adi Roche spoke about the need to look towards a "recovery" phase and this is the focus of all the UN programs in the region.

People living in contaminated areas cannot buy clean food if they don't have an income. So the idea is to invest in infrastructure to build up the economy. The government of Belarus is offering tax incentives to companies willing to invest in the Gomel region. They've put up a web site touting the great real estate deals. The ads don't mention anything about cesium-137.

People like Danilo Vezhichanin, the mayor of Yelna, aren't particularly interested in development. They'd really like to go back living the way they did before the accident. He said that the he and the people in the village just want clean land and to grow their food. But that isn't going to happen, so like it or not, they will be "developed."

The United Nations Development Program (UNDP) issued a report in 2002 about the problems in Chernobyl and the international obligations to the people there. The report emphasized the idea that people in the area are suffering mainly from the psychological and social shock of being uprooted from their homes and communities.

The Chernobyl Forum report issued in 2005 was a direct offshoot of the 2002 UNDP report. Pavlo Zamostyan is a coordinator for the UNDP in Ukraine. He said that one of the goals of the UNDP is to work with the governments in the affected areas to come up with realistic policies. He described the current system as an economic black hole that the government couldn't realistically keep. He says that in 1991 many people were told they would be resettled but after the fall of the Soviet Union that just wasn't possible. People who did get new housing got only that. Often the houses were in places with no infrastructure, no roads, schools, markets or jobs.

The current policies in Ukraine were drawn up in the idealistic period after the fall of the Soviet Union. Many Ukrainians saw Chernobyl and its aftermath as symbolic of the problems with the Soviet Union in general. Giving generous compensation to Chernobyl victims was a way to right some of the wrongs they suffered under Soviet rule.

The idealists were soon confronted with the reality of the situation. There simply wasn't enough money to keep the health and social benefits going.

The same realization was dawning in Belarus and Russia. That's when the IAEA, the UN and WHO came up with the policies outlined in the 2002 report and further detailed in the 2005 report.

A principal difference between the old policies and the ones being promoted by the UNDP is that the community must provide a substantial part of the capital for the projects. Unlike humanitarian aid, which provides 100% of the funding, the communities must come up with 60% to 70% of the resources. That can be in-kind contributions in the form of donated work time or it can be cash collected from local government and businesses. The idea is to get people out of a victim mentality.

Zamostyan is optimistic about how the program will work, "If there will be good social economic conditions maybe people they will start to think that, well, their health problems are similar to those in other territories and maybe they can just have better access to normal medicine."

The President of Ukraine, Victor Yushchenko, outlined a plan for the future of Chernobyl in December 2005. He suggested that the area might be used to reprocess nuclear fuel for Ukrainian power plants and as a storage facility for other countries nuclear waste. The idea horrified others in the government. Within a few days of Yushchenko's announcement, the Ukrainian Parliament Speaker Vladimir Litvin dismissed the idea, "We should drop this subject and set the minds of those who suffered from the planetary catastrophe at rest." He suggested anyone who wanted to turn Ukraine into a dumpsite should "experiment in their own orchards first."

While Litvin didn't like the idea of bringing in more waste, he liked the idea of bringing in tourists. In a page out of the "make lemonade out of lemons" school of public policy he suggested that the site could be used for "extreme tourism."

One of the ironies of the whole Chernobyl situation is that Belarus, which has so much of the radioactive fallout, doesn't have any nuclear power plants. There are no plans to build nuclear waste reprocessing or dump facilities in the contaminated areas in Belarus. They don't even like the idea that a nuclear waste facility in Lithuania will be close to the border of Belarus.

In 2005, Lithuania announced plans to build a nuclear waste facility about 700 meters from the Byelorussian border. Minsk could only retaliate with pig farms. They threatened to build two pig farms near the Lithuanian border. The Lithuanians were appalled by the idea because waste from about 216,000 pigs could flow down the Neman River to a popular spa resort in southwest Lithuania.

The Lithuanian Prime Minister, Algirdas Brazauskas was against these plans.

It's easy to understand why Belarus would not want any more radioactive material nearby. In 2006 the government of Belarus estimated that during the next 5 years they would spend $1.5 billion on Chernobyl related problems. It's a substantial amount of money for a country still struggling with its Communist past.

An estimated 23 per cent of the Belarusian territory is severely contaminated with caesium-137. About every fifth inhabitant of Belarus lives in a contaminated area. Dr. Vasiliy Nesterneko founded the organization "Belrad" and has been studying the radiation in Belarus since 1986. His assessment is not optimistic, "Belarus will never overcome the Chernobyl accident effects if using only its own potential, - the damage caused by the accident measures $270 billion, more than 38 national budgets."

The power plant was built by the USSR and because that country no longer exists, Belarus has no recourse for reparations. Russia, the legal successor to the Soviet Union, maintains that it has no obligation to compensate other countries for the damages. Ukraine, where the plant is located, also denies any responsibility.

In 2005, the Federation Council of Russia ratified the Vienna Convention on Civil Liability for Nuclear Damage. The Convention regulates the procedure for paying compensation for damage from accidents at civilian nuclear power plants. Many believe that by signing on to this treaty, Russia should be liable for compensation to Chernobyl victims. But after signing on to the treaty Sergei Antipov, deputy head of the Federal Agency for Atomic Energy, said that Russia was not liable because the disaster took place before they signed on to the treaty and it occurred on the territory of another state.

Even though Russia refuses to take financial responsibility for the consequences of Chernobyl in other countries, it isn't free from all the problems. The Bryansk Oblast of Russia, which is about 66 miles northeast of Chernobyl, was heavily contaminated by Chernobyl's fallout. The area is experiencing the same sorts of problems seen in heavily contaminated regions of Ukraine and Belarus.

The debate over Chernobyl will probably fade away before the radiation does. Already people who were born after the accident have no idea that it happened. Ask a person born in the late 1980's about Chernobyl and they'll probably have no idea what you're talking about. Once people stop asking questions about the accident, nothing will be found.

It's only when researchers know to look for statistical anomalies and know to look for radioactive fallout that the effects can be detected. It's disturbing to find that a well-respected institution like Harvard University could publish the report "Autopsy on an Empire: Understanding Mortality in Russia and the Former Soviet Union" and completely ignore Chernobyl. It may be that the puzzling drop in life expectancy has nothing to do with Chernobyl. But when the researchers don't even discuss it then it's as if it never happened.

Chernobyl and Three Mile Island brought the nuclear power industry to a grinding halt. In the United States no reactors have been brought on line since 1996. From 1996 to 2003 there wasn't even a single application submitted to the Department of Energy for a license to start construction. The same trend was seen all over the world. When Hans Blix stepped down as Secretary-General of International Atomic Energy Agency (IAEA)in 2005 he reportedly said, "Global nuclear power is coming to a standstill."

But that may not be the case for long. In the United States the 2005 Energy Policy Act provides funds for nuclear energy research and development. It includes incentives for building new reactors, loan guarantees, production tax credits, and investment protection for delays beyond the builder's control. When George W. Bush signed the act into law in August of 2005 he said, "We will start building nuclear power plants again by the end of this decade."

The Department of Energy is already helping energy companies get permits for three new nuclear reactors by partially funding the process. The new reactors would be built at current plant sites; the North Anna power station in Virginia, Clinton in Illinois, and Grand Gulf in Mississippi.

Even without any new reactors, the United States has, by far, more reactors than any other country. As of December 31, 2005 there were 449 nuclear power plants in operation worldwide. The United States had the most with 104. France had the next highest number but they had only 59.

The rest of the world is also looking at building new nuclear power plants. At the end of 2005, India had 8 reactors under construction. China's National Development and Reform Commission has proposed a long-term plan for nuclear power development that calls for building one nuclear power station every year for 16 years.

Every single one of those nuclear power plants will have to be run safely for decades, whether it is in Finland, Pakistan or China. The engineers who design and build those plants are confident that they've thought of every single combination of events that could lead to a catastrophic accident. The construction companies will be certain that everything is built to specifications. The authorities that will operate them will ensure that all workers are free to discuss any safety issues and safety will always be put before profit and productivity.

Regardless of what administration is in charge of the government, the reactors will have to be run with nothing but the highest standards of safety. It's a lot to ask but it only takes one accident at one reactor anywhere in the world to have effects that last for generations.

It may be that the technology exists to build nuclear power plants that don't leak or explode. We may even be able to protect them from terrorist attacks and figure out what to do with all the waste. The question is how much do you want to bet on that? How many cancers are acceptable? How many birth defects? How many children are you willing to sacrifice?

Every nuclear power plant is a new source of radioactive material. The waste it produces will remain dangerous for thousands of years. It can affect the genetic make-up of people for far longer than the recorded history of human civilization. Long after our cities have fallen into disuse, after our languages are no longer spoken and every memory of this civilization is gone, we will leave a legacy of disease and death to our descendants. This can't possibly be worth the price.

How The Area Is Supervised Today

Shortly after the nuclear power plant disaster happened and the people were evacuated, the government declared an exclusion zone of about 30 km around the power plant. Over time the zone was expanded several times due to new fallout coming out of the plant. Today the zone covers around 2.600 km2 in Ukraine and Belarus.

The main purpose of the exclusion zone is to prevent people from entering areas with high radiation and to prevent the spread of dangerous radioactive elements.

Access is only allowed for people working in the zone and at the power plant and during controlled organized tours.

Besides the cities of Chernobyl and Pripyat the zone is full of scattered little villages and farms. Most of the remote farms and smaller villages were evacuated at a later stage depending on the state of contamination.

Parked inside the zone are also many vehicles, left behind after the cleanup operations. These vehicles couldn't be taken out of the zone because they were often highly contaminated during the cleanup and since the area is already contaminated it was best to leave them inside the zone.

Because the zone is so empty, it is also fairly easy for people to go into the zone and take away building materials and other usable objects.

This is a big problem because they are potentially contaminated and therefor spreading the radioactivity outside the zone.

Flora and fauna are doing extremely well in the zone.

Forests are now where humans once lived and soon it will be hard to spot any human activity in many of the more remote areas.

Animal life is doing good as well. Wolves, wild boar, deer, badger populations have multiplied over the past years.

One other thing that stands out while being in the zone is that time stopped in 1986. After a while you notice that you are still in the Soviet Union. All the propaganda from the Soviet times is still there and the many posters and billboards of the great leaders are still prominently visible.

The toys and household items scattered around the buildings and houses give a good insight of how people lived.

Everyone had very similar wall decoration, furniture, etc.

Life here wasn't about having the latest trends; life was about having functional items.

Which Countries Have Suffered Other Nuclear Accidents

On March 28, 1979, the greatest nuclear disaster in the US unfolded at the Three Mile Island Nuclear Generating Station in Pennsylvania. A valve stuck in the open position allowed cooling water from one reactor to escape. More than 32,000 gallons (121,000 liters) of water–enough to fill two swimming pools–bubbled off. The reactor shut down as it should, but there was confusion in the control room. Faulty sensors meant operators couldn't tell what the problem was. The workers thought the

cooling system had plenty of water, so they shut down the emergency cooling system. That system kept coolant flowing to the core. It was hours before workers turned the system back on.

In the meantime, half the core melted

Ohio

New York

Pennsylvania

West Virginia

Virginia

New Jersey

Delaware

Maryland

Connecticut

Three Mile Island

The Three Mile Island Nuclear Power Plant is about 10 miles (16 kilometers) south
of Harrisburg, Pennsylvania, the state's capital city.

EXPLORING PRIPYAT

We're approaching the hotel. I pass some harrowing graffiti: the black silhouettes of children playing, painted on the hotel's restaurant walls. You can see for miles up here. Chernobyl sits on the horizon behind abandoned homes, while the ferris wheel apex crests a carpet of trees 150 meters away. As the others busy themselves taking photos from the roof, I separate from them and head towards it. Walking outside on my own for the first time, I gaze across the overgrown square with its cracked concrete and recall decades-old photographs of sunny days, pristine rose bushes, parades and smiling faces. It's so lonely here in the present. I'm a solitary sort of person and have fantasized countless times about how extraordinary it would be to be the last person on Earth, to go anywhere and do anything I wanted with absolute freedom. Post-apocalypse stories, in particular, have always held sway over me. How ironic that, now I'm experiencing a sliver of that imagined existence, it unsettles me so.

It's a strange feeling when you first see something so familiar from photographs with your own eyes, like visiting the Eiffel Tower or the Pyramids, but familiarity does not preclude awe. You know all the major details, the colors and shapes, but there's so much you never noticed before. There's vital context too: you see everything around it, the geography, and distant things you hadn't expected to see from that particular spot. Near the ferris wheel, which was never formally used, as it'd been due to open as part of the May 1st festivities, are the famous dodgems. The exposed metal base is one of the more radioactive parts of the city, but the cars themselves are in decent shape, all things considered. I saw a great photograph of them once and I try to compose my own, but find myself distracted by thoughts of disappointed evacuee children back on May Day, 1986.

I suddenly realize I've been on my own for half an hour. I had expected Danny and the others to join me minutes after I left, but there's neither sign nor sound of anyone. Maybe they weren't heading in this direction until later? Did I actually tell anyone where I was going?

I set off back towards the hotel, glancing up at the roof I left them atop, but not seeing any familiar faces gazing over the edge. Perhaps they headed into the sports building from earlier. Did Liquidators drain the water, or has it evaporated over time? Regardless, there's nobody here. Have they walked off and left me? A square canvas painting, my height, with a celebratory 'CCCP 60' emblazoned in bold white text against the traditional Soviet blood-red, sits propped against a pillar in the building's large entranceway. It turns out that the sports building is the rear of Pripyat's Palace of Culture, one of the city's most recognizable and central landmarks.

The stage area roof is higher than everything else in the Palace of Culture, enabling light racks to hang high up out of view of the crowd. Those same lights are now slumped across the stage.

A drastic plan to seal Fukushima Daiichi off from the surrounding earth, to stop contaminated water from leaking into the sea, was approved and the required machinery built. The joint TEPCO and government effort involves freezing the ground using 1568 pipes in a colossal wall 30 meters deep. Critics of the plan pointed out that cost and feasibility issues were not properly thought through, but the government pushed ahead with it anyway. An initial attempt to freeze the earth ended in embarrassing failure in 2014 when TEPCO couldn't get the temperature as low as required, even after adding ten tons of ice into the mix. Freezing has since repeatedly failed to contain all of the water, despite pumping around $325 million of public funds into the project by March 2018.

The machine never worked properly and only filtered a total of 77,000 tons of water, instead of the 300,000 it was intended to process every day before it was abandoned. The leaking storage tanks mentioned above somehow cost $135 million, all of which are being replaced.281

Nuclear power had been experiencing something of a renaissance before the Fukushima disaster, with the world appearing to finally move on from Chernobyl. A fresh emergency brought old fears back to the surface, causing many countries to review their nuclear policies. For their part, Japan immediately shut down all 48 of its remaining nuclear reactors after the accident in 2011, though it has since reactivated a select few. Nuclear power remains a divisive topic in the country, with strong public opposition. Germany, another major user of nuclear energy, followed suit and announced plans to begin decommissioning all of its plants, along with Sweden and Italy. Even France, famous for relying on nuclear power for about 75% of its electricity, has shied away from atomic energy and plans to reduce its reliance on nuclear energy in the next few decades.

The reactor has suffered from multiple delays and was originally supposed to be finished in 2012, then 2017, but as of June 2019 is planned to enter service in 2020. It can operate for four months without any human control and has been designed to last 100 years - triple the usual lifespan. Competing renewable energy technologies like wind and solar are improving all the time and may be a viable alternative to coal, oil and nuclear fuels in a few decades, but for the time being nuclear power seems like our only realistic chance of creating clean energy on a global scale. Let's hope that those with the power and money to build and run them have learned to put safety first.

Our sharply increased awareness of each other has also seemed to have caused a great increase in the problems facing humanity. I say, "seemed," since the problems, for the most part, have existed for some time, although their intensity has deepened greatly in recent years. We have become sensitized to problems of poverty, pollution, popula-tion, and peace on a global scale and have come to realize that solutions likewise must be planetary in scope if they are to be effective.

At the same time, we are presented with opportunities that are even more significant than the problems. Buckminster Fuller states that "Humanity is going through its final examination to see if it can qualify for its universe function."[1] Whether it is humanity's final examination, it is without doubt its most important in all of human history. On the one hand, human survival is at stake. On the other, so is humanity's new role as conscious steward of evolution on this planet. Four billion years of earth's history have brought forth a species that is not only creative, self-actualizing, and capable of compassion and empathy, but also one that is globe-girdling and globe-dominant. The immediate corollary of course is responsibility. We have become responsible for the entire earthly domain. The open question before us is, Are we wise enough to execute that responsibility?

The execution of global responsibility depends upon the grounding and demonstration of appropriate human characteristics and values. These characteristics and values are no different from those required in good relations and effective collaboration among individuals. The same are now required from our largest human collectives—the nation-states.

What are some of these values and characteristics? There is nothing new about the homely values of goodwill, caring, and sharing; of making space in which others can grow and become; of helping to create environments that maximize the expression of human potential, alleviate suffering, allow and protect individual-ity, take responsibility for group good, limit and then transmute aggression, curb and moderate acquisitiveness and the violent and selfish pursuit of individual desires. These are values that have been found somehow not only to be socially necessary and desirable but also intrinsic in the evolving pattern of human growth as the capacity for decentralization and for identification with the hopes, needs, and wishes of others develops. They are necessary for the emergence of community. They are equally necessary for the emergence of world community. There must be some minimum threshold expression of these values on a global scale before humanity can integrate as the essential subjective unity it is and successfully manage planetary affairs.

Government Declarations

One must quickly add and acknowledge that these Covenants are often, if not usually, still honored in the breach; that even the United States has not yet ratified them; that codification is one thing and implementation a much more difficult step. For some countries and people particular human rights seem more signifi-cant and timely than others: The West is preoccupied with civil and political rights—rights to individual freedom and liberty. The south—the Third World, is much more concerned at present with rights to food, health care, education, clothing, and housing. These differences in current emphases must be respected and acknowledged while we insist that no human right is secure in the absence of the expression of others.

We also have before us on the international agenda the dim outlines of a value pattern for a global economic system. These have been set forth in the "Declaration on the New International Economic Order "and a "Charter of Economic Rights and Duties" adopted by the United Nations. These are certainly not perfect documents, and they brought sharp opposition from the biggest industrialized countries on a few points—for example, compensa-tion for expropriated assets. Nevertheless, we are now on the way to defining and then to implementing the economic obligations of peoples for each other.

Ecological values have been institutionalized in a new UN Environment Program agency; those concerning health, educa-tion, and welfare, in the World Health Organization, UNESCO, and the UN Development Program.

There has thus been a successful transfer to the global level of the principles of commonly held human values. I say the "princi-ples" since these agencies are still embryonic, weak, and poorly supported. However, the value transfer in itself is not only successful but immensely significant. It represents among other things a high degree of tacit international agreement on a defini-tion of a human being—a surprising and unplanned development.

Set against the translation of humane values to the global level must be the pursuit by many nations of goals that are incompati-ble with world community well-being. We must also put on the deficit side of the equation collective aberrations and short-comings of peoples-as-nations, aberrations that have strong ana-logues in individuals human behavior. Thus, for example, one of the main fuels of the suicidal nuclear arms race is erroneous threat perception — often based on "projection" of one's own notions or motivations. Nations tend to accept a "worst possible case" definition of the intentions of others. This leads to actions on their part, such as deploying a new and more destructive weapons system, which result in reciprocation and confirmation by the "enemy" state or states. Thus a self-fulfilling prophecy is produced, gravely decreasing the security of all concerned.

Another easily recognized psychological condition affecting the relations among states (as it also does among individuals) concerns the problem of communication. States, like individuals, tend to screen and to reject information at variance with that which they already have. They tend to accept information that reinforces their concepts of what is "true" and to reject contrary information as patently "untrue." This accounts to a large extent for the tardiness with which national attitudes and policies shift and for their almost continually lagging far behind newly perceived realities.

The question arises why the largest states often appear to have the most difficult time in perceiving and accommodating to change or in directing themselves to the perceived general inter-est of the world community. I have speculated that one factor might simply be mass. A large mass is less mobile—in this case, in opinion formation. It may also tend to be more ego-centered and self-involved. A large state traditionally cares little about the concerns and needs of the small. It may, once it does swing into action, act overwhelmingly and immovably on the basis of filtered information, half-digested "intelligence," and false assumptions. In other respects, too, states resemble individual humans. No more than individuals are states unitary or wholly integrated. Just as the human is often a "divided house" of conflicting needs, aspirations, wishes, drives, and motivations, so is a nation. Some of these divided dominions can be identified as political parties, ethnic groups, and leadership groups from various economic strata, academia, large and powerful business and labor groups, and so on. These constitute, in a sense, some of the "subpersonali-ties"of nations. Of ten national objectives and policies necessarily reflect a "lowest common denominator" resulting from the goals and desires of these constituent dominions. Sometimes a forceful and well-motivated leader can challenge these groups to their best motivational potentialities.

The Chernobyl explosion was not the first nuclear disaster in the USSR. The Soviet government reacted the way it had in the past. They classified all the information about the accident. Everyone involved was sworn to secrecy. After all, it worked so well with Chelyabinsk.

For almost twenty years the outside world knew nothing about an incident in the Ural Mountains near the town of Chelyabinsk where buried atomic waste exploded in 1958. The story didn't come to light until November of 1976 when a dissident biochemist, Dr. Medvedev, wrote about it in the British Weekly New Scientist. Even then, the government denied that anything had happened even though tens of thousands of people were suffering from radiation related illnesses.

But the earlier accidents were contained within the Soviet Union. Chernobyl's radiation blew into Western Europe. The first person outside the USSR to detect the radioactive cloud was at a monitoring station in Poland. When the operator saw the high levels of radiation he assumed the equipment was faulty and ignored the readings.

On April 27th, workers at a nuclear power station in Sweden discovered radioactive particles on their clothes. The Swedes knew their equipment was working properly and they tried to find the source. At first they thought it was a leak from the Forsmark atomic power station on the Baltic Coast just north of Stockholm. Emergency personnel arrived at the power stations and took measurements around the power plants. Radiation levels inside the plant were lower than outside. On April 28th, Swedish radio announced that a nuclear accident had happened somewhere inside the Soviet Union.

As the radioactive cloud spread, people all over Europe were told not to eat fresh fruits and vegetables. Children were to be kept indoors. Even though the accident happened thousands of miles away, it couldn't be ignored. Its consequences would haunt parts of Europe for decades. People living the closest to the land were affected the most.

On April 26th, a heavy rain fell on the mountains in Northern Sweden. Karin Baer, a Sámi woman, had just arrived in the mountains, bringing her reindeer to the summer pastures. She remembers hearing the rain on the hard roof of her shelter in the early morning and thinking it was a good sign. The snow would melt, making it easier for the reindeer to find food.

The Sámi people, sometimes knows as Laplanders, have lived in what is now the northern territories of Sweden, Finland and Norway for thousands of years. Up until the middle ages, they survived by hunting the reindeer that thrived in huge herds. During the middle ages, they began to somewhat tame the reindeer and to mark specific animals as belonging to specific people.

They kept an eye on the herds, killed off the predators like wolves and bears. They developed a way of life that revolved around moving with the reindeer. Anything that affects the reindeer, affects the Sami.

When Baer heard radio news reports about high levels of radiation at the monitoring station she knew they were in for trouble. She knew all about Becquerel's and curies from the Russian bomb tests in the 60's when they were doused with radiation. Everyone knew that the radiation concentrated in the lichen that the reindeer ate.

Elna Fjelstrom and her husband didn't have a radio but they knew something was wrong. As they worked outside in the rain Fjelstrom's skin began to itch and peel off in strips. A few days later, other Sami's came up into the mountains and told them about the accident.

Her skin healed and they didn't think much about the accident until that fall. For the first summer after the accident no one checked radiation levels. But because the Sámi had dealt with radiation problems in the past, they expected some problems. They didn't understand the size of the disaster until it was time for the fall slaughter.

The first sign of trouble was when the government told them that the children shouldn't be around when the animals were killed. Health workers worried about the children breathing in contaminated steam rising from the reindeer's bodies when the bellies were slit open. It wasn't a small matter. Sámi children are always part of the round up and slaughter. It's how they learn the trade.

The children aren't just by-standers. They have a stake in the whole process. Often their parents will register a set of ear markings for their children. During calving season, some of the parent's calves are marked for the children. It takes decades to build up enough reindeer for a person to be able to make a living at herding and this is a way for parents to pass on the heritage.

Just before the fall slaughter Swedish scientists showed up with devices to measure radiation levels in the animals. Not a single reindeer was edible.

The old people cried when the reindeer were buried. Even though the government promised them that they would be compensated for all the reindeer they couldn't eat, there was more than just money involved. People watched as an entire lifetime of careful husbandry was destroyed. It wasn't just meat being thrown away. When the Sámi say that they were afraid that their way of life would die out because of Chernobyl, they weren't speaking metaphorically. Legally Sámi are defined by their reindeer not their parentage. Even if you were born into a Sámi family, you must also own and take care of reindeer to get the right to use the land, fish and timber in Sámi territory. If they lose the reindeer, they lose all legal recognition of their people.

They've learned to live with the radiation with lots of help from the Swedish government. It's just one more adjustment to the traditional ways. They still follow the reindeer but now they move the herds using helicopters, motorcycles and snowmobiles. The fall round up is still an amazing sight.
We arrived near Klimpfjall, a little town in Northern Sweden, as the Sámi were preparing for the fall slaughter. We drove about 10 km outside of town until a tall wire fence blocked the road. No

one could use the road until the reindeer crossed. Sámi dressed in Gortex jackets and boots waited near cars and pick up trucks parked by the side of the road. In hushed tones, they told us that the reindeer would be arriving soon. A woman watched the horizon with a pair of binoculars. We heard the helicopter before we could see it. A two-seater, blue helicopter rose just over the tree line of a nearby hill. The woman with the binoculars told us to move a few hundred meters away from the fence and crouch down so that the reindeer wouldn't see us. The first few reindeer bounded over the rise. Then the entire herd spilled down the hillside. It looked like a liquid mass of soft brown and tan fur bristling with antlers. The fence across the road formed a narrow passage where the reindeer pressed against each other as they ran from the helicopter. The low flying helicopter drove thousands of reindeer over the hill and into a nearby valley.

Once they were across the road, they fanned out into the valley below the road. The Samis moved the fence to clear the road and the reindeer couldn't move back up the hill. The helicopter flew off.

We stood by the road for a long time. The reindeer are about the same size as the white-tailed deer on the East Coast of the US. Occasionally, two males with huge curving antlers locked horns to form a thorny ball between them. Then they pushed against each other, clacking their antlers together violently as they twisted their heads. Both males and females have velvet-sheathed antlers but it hung in ragged strips on the combative ones. The castrated males have oversized racks that make them look unbalanced.

Their fur coats range from dusty brown to ermine. They make a sound something like a cough and a grunt. Some of them were wearing square cowbells that clanged unmusically. Others sported bright, hunter-orange collars with reflective strips which make them visible to car headlights at night. Ear nicks and splits identify their ownership. As we approached the fence, they moved away from us. They are skittish about being near people. I could understand why when we saw the next phase in the process.

The Sámi used dirt bikes and ATV's to move the herd into a circular area enclosed by a slat-wood fence. The reindeer run around and around looking for a way to escape. At first they form a tight circle in the center of the arena. Eventually they spread out a little but they never stop running.

Once the deer have dispersed it was safe to go inside the fence. It was very strange to stand in the middle of this stampede of deer but they avoid people by parting like a school of fish around anyone standing in their midst. They never come closer than about a meter, which can be scary until you see a Sámi toddler calmly holding his mother's hand while these beasts with long, pointy antlers thunder right past them.

Which Countries Have Given Up Nuclear Power Plants?

Nations differ widely in their degree of integration and self-actualization. Some are much more value-oriented planetary citizens than others. The voting records of some nations at the UN are consistently more community oriented, more expressive of responsibility and concern than others. A world community oriented scoring system that I have applied to voting at the UN General Assembly for a number of years consistently shows Australia, New Zealand, Yugoslavia, Singapore, Canada, the Nordic Group, Colombia, Venezuela, Nigeria, Ghana, and Sri Lanka among the "high" scores.2

It is interesting to observe that quite often countries maintain a forthright and community-minded international policy over a considerable time even when their internal situations contain strife and repression.

Perhaps the Scandinavian countries have an easier time than most in developing a constructive attitude toward world affairs. They have homogenous, relatively small populations whose basic needs are met and no serious internal problems. They have developed over time a level of social concern that has now extended as a natural development to the world community as well. Sweden and Norway, for example, are among the few countries to approach or reach the UN goal for official develop-ment aid.

National personalities are very recognizable if one has the advantage of an international or non-national perspective as Planetary Citizens should. There are new, fresh nations, with idealism and growing capacities—often represented by their fin-est people; there are old and cynical nations with great skills, often misapplied; there are late adolescents, like the U.S. and U.S.S.R., with very different and also very difficult (and not equal) motiva-tional and adjustment problems.

Unfortunately, very little study has been given to the psycho-logical characteristics and well-being of nations as such. Political figures are not given to what they would consider irrelevancies such as psychological speculations. Psychologists and psychia-trists have largely ignored the field partly out of fear of criticism by their colleagues. Some exceptions have been Bryant Wedge with his Institute for the Study of National Behavior; Professor Jerome Frank, formerly of Johns Hopkins; Professor Herbert Kelman of Harvard University; and Professor Charles Osgood, of the University of Illinois at Urbana.

It is unfortunate that there has been so little work done, since a clear understanding of the psychological states and stages of nations-as-entities can be crucial to understanding behavior and to initiating successful conflict-resolution measures. Serious per-sonality, motivational, and perception differences between or among nations today can mean the expunging of the human race and the end of the human experiment on earth. I earnestly hope for the development of a new psychology of groups and collectives—most particularly of nation-states. As the concept of world community continues to develop, there is bound to be a growing recognition of the personality characteristics and prob-lems of the primary actors in that community—the nation-states.

In regard to Western industrialized countries in particular, are there any indications of a hopeful nature suggesting a maturing of values and attitudes toward the rest of the community of which they are a part? There is one particular and quite new indication I should like to note. A trend away from consumerism, acquisition, and satisfaction of material fancies at all costs has become suffi-ciently apparent in the United State to have been the subject of an important study by Stanford Research Institute, a study directed to United States businessmen.

The authors find a significant proportion of the Ameri-can population, as well as that of some other Western countries, is beginning to opt for voluntary simplicity rather than for what some feel is an endless and mindless pursuit of material ends.

The trend toward voluntary simplicity has many implications, especially in terms of maturity and in terms of a deeper emphasis on inner growth. Elgin found a close correlation between interest in the new psychologies and the practice of some form of medita-tion and an opting for voluntary simplicity. For our purposes such a trend within Western industrialized nations can become part of a significant accommodation between the wealthy, northern "have" states, and the "have not" southern countries. Such a value trend can smooth the way to new accommodations and under-standings between these two groups. Thus we are privileged to witness important and significant value formations in process.

Nor is participation in the world community ultimately anti-thetical to, for example, United

States traditions or concerns. The U.S. concern and penchant for effective management is quite naturally going to thrust it into the forefront of efforts at satisfac-tory and cooperative planetary management. A national accep-tance of consensual and democratic decision making could allow the U.S. to take leadership in the introduction of these approaches in international forums.

Other international personalities are more intractable and diffi-cult, holding per solutions for world problems in simplistic and outmoded ideological terms, while they hold their populations to regimes of sharply limited self-expression and, in a sense, to "enforced simplicity." However, if the world will persevere in a move toward cooperative and constructive directions, hold-out nations will have little choice but to fall back on "me-too-ism." This has, in fact, frequently occurred in UN forums, even where their initial opposition was adamant.

One must say, however, that in general world political leaders have tended to tramp on and exacerbate the raw nerves of the psychoses that nations suffer, an approach that has certainly not been found useful in treatment of individuals. They tend to respond in kind or to react in righteous indignation to the provo-cations of others rather than to seek the avenues that will allay suspicions, fear, and doubts and allow for the possibility of changes in attitude in morose nations-personalities. In this respect, we await our clinicians of nations. The United Nations itself is a fascinating and very promising psychosocial arena. The major powers are, of course, a little disenchanted with the UN at present. This is primarily because the tight control that some of them used to exercise over the UN is no longer available to them now that the UN has become a universal instrument, and big and small nations must now compete on more equal terms for the exercise of nation-state policies. However, the intensity and continuous nature of communication at the UN provides multiple opportunities for dealing with the difficulties of interpersonal national relationships. To speak anecdotally, I have on more than one occasion been witness to skilled therapeutic handling of interpersonal problems and national policy questions by UN diplomats and secretariat people dealing with the suspicions, fears, and unreasonable doubts of diplomats from "closed-in" countries. I can say quite definitely that in many cases nations do tend to choose UN representatives who epitomize personally the condensed and specific psychological sets, difficulties, and cultu-ral qualities of the countries they represent.

There are, of course, many levels to UN proceedings. There is the level of official government policy. There is a level of personal concerns and integrity that sometime enters in; and then, too, "national" and "personal" are not always the same. There are the unstated policies of the UN secretariat, which as a neuter civil servant, is expected to have none. In fact, however, the UN secretariat is a reservoir of both skills and vision and can often succeed in placing before nations the next steps needed for the common good of the world community.

There are further levels. There is a level born of the simple and concrete fact that elements from all parts of humanity are gathered there in one spot: a kind of sensitive switchboard to all of humanity—unique in history, unique in human experience. Approaching universality at present with 51 member states, the UN is one of the few locations on earth where some synthetic feeling of what humanity itself is can be grasped, felt, sensed, and experienced. I should like to assure you that quite apart from the conflicts and problems of nations, this is an unique and elevating experience. Willy-nilly deeper levels of experience and meaning emerge at the United Nations because of this fact. The unex-pected impact is often remarked upon by short-term delegates or by visitors. Some major personal reformations in attitude have been accomplished by exposure to this unique total-human environment. As is so often the case, that which appears threaten-ing is that which has not been experienced. When one comes to know firsthand the aspirations and goals of others, the unknown loses its fear.

The UN provides a unique setting in which it is possible to step outside of any particular national framework and to view the world pattern with a new and detached perspective. The prag-matic, practical idealism of many members of delegations, of unsung civil servants of the secretariat, and of nongovernmental groups ("NGO's") close to the UN that have achieved this pers-pective constitutes a considerable "natural resource" upon which the world is only beginning to draw. There exists in and around the UN an informal but tangible network of such people that Planetary Citizens has dubbed a "humanity underground" concerned with the good of the community. They represent a group of "honest brokers" in the harmonization of competing or conflicting national goals and hold a clearer vision of world community needs than is available to persons bound by partisan national views.

The world public has yet to take the measure of the two secretaries-general, Dag Hammarskjold and U Thant, who personified this role and this view, and who, incidentally, were making concerted personal efforts at inner growth. U Thant was little known or appreciated by Westerners; his memoirs, View from the UN give some profound insights but are not as revealing of his intense inner life as was Dag Hammarskjold's spiritual diary Markings. Another reflective element has been added in recent years to the UN by Sri Chinnoy, who heads the UN Meditation Group, meeting twice weekly with the participation of secretariat members, delegates, and NGO's.

Thus the character of the UN may be quite different from newspaper accounts of the political confrontations, which remain, of course, a consistent feature of its proceedings. The UN is gaining in centrality and importance even while it is being decried, because it represents an historical imperative. A plane-tary center has become essential to the conduct of human affairs. In the UN a planetary core has been established, a point for focusing human synthesis. A new awareness is emerging in the world; and humanity itself, like Rip Van Winkle, is waking up, collectively to take a charge of its own affairs and of spaceship earth—and none too soon.

Mankind's consciousness now must stretch to a new level; our inclusiveness and identification must now include the entire human community—the fellow beings of our species—and that species must share an appreciation and reverence for all planetary life, if indeed it wishes to survive and persist on this planet.

The nations of mankind are approaching a new threshold beyond which they will be integrated as components of a new and self-aware organism—global humanity. A virtual Copernican revolution in awareness confronts us, with the opportunity and the necessity for redefining our relations with each other within the species; for defining for the first time (because it is only now historically possible) the relation of the unit to the entirety; for redefining our relations with the planet, and in a sense, with the cosmos.

Such redefinitions, or their lack, are bound to have significant impact on the state of the mental health and well-being of nations and individuals. In the same sense that in the past, family, tribe, community, state, and nation have represented new aspects of ourselves that required new identifications, and incorporation into our mental and feeling world, now another step is required. The borders of the nation-state as the ultimate and final boundary of identification, of "self" and repository of loyalty are becoming less sharp now, and a sense of loss, frustration, and confusion will ensue for the person who does not take the step to the final planetary loyalty and identification. There can be no more an "out group" of the human family if survival is to be assured. We must enlarge our capacities for acceptance to include all peoples. Thus it seems quite clear that human health and wholeness will depend henceforth, in addition to other factors, in crossing this new threshold to planetary awareness. We might even say that in a very real sense humankind, facing the crisis of becoming whole, faces the possibility of being healed.

Science advances through a succession of "paradigms, "or frames of reference, which are often mutually irreconcilable, according to Thomas S. Khun's familiar description. The term "paradigm," widely adopted in the social sciences, has become almost common parlance.

In the process, however, it has lost some of its meaning. People now speak of paradigms wherever two or more concepts show any semblance of systematic coherence. They also speak as if a babble of different paradigms—"Marxist, realist traditionalist, peace research, feminist, and behavior list," to quote one recent catalogue—could exist side by side. Everyone has the right to use the term as he or she wishes, but this is not the sense in which Kuhn intended it.

Kuhn was concerned to show that even in science, where the raw data of the scientist's observation can be limited and con-trolled, they would be unmanageably complex without some previously agreed upon frame of reference. He showed that such a frame of reference involves preset categories which soon become deeply rooted in the perceptual and thinking process, and that, until "paradigm breakdown" and "paradigm shift" occur, it is widely if not universally shared throughout the community of discourse. One can be a Marxist and a feminist; one cannot believe in a Ptolemaic and Copernican universe.

This is not to say that people do not use preset categories in life, as they do in sciences; on the contrary. As William James pointed out, without selective perception to screen out the majority of our sense data and our own thoughts, life, even more than science, would be past coping with. As Hazlitt put it, "Without the aid of prejudice and custom I should not be able to make my way across the room."

Yet in life, even more than in science, paradigms can be disas-trous failures. Or to speak historically, paradigms that have provided a useful system of selective perception, evaluation and decision for a time can become worse than useless encumbrances and require wholesale replacement for society to advance.

That is precisely the position we have now reached with the prevailing attitudes about militarism and war. Many peace researchers today would agree that the achievement of stable peace conditions in the world will require a paradigm shift. They may not all realize, however, that the shift required will not be that spoken of loosely by the social scientists, but the much deeper and broadly accepted change of vision described by Kuhn.

What is needed is not just a change of opinion, like the oscilla-tions in public legitimacy in the United States accorded to the Vietnam War, but a permanent shift in how we view the world: how we gauge hostility, what we think of to do about it—almost a shift in what we perceive as real. Aristotle actually "saw" con-strained fall in a stone swinging back and forth on the end of a string, Galileo saw the glimmerings of momentum in exactly the same phenomenon. As long as people feel comfortable when they make threats to other nations of people whom they regard as enemies and fail to perceive that those "enemies" will respond by making counter-threats (just as they themselves do), it will not be possible to abolish war.

It is true that individual wars can be aborted by better diplo-macy. It is also true that the tendency of nations to get into dangerous confrontations can be mitigated by the institution of more rational political and social systems within them. But if we want to eliminate the root cause of war—and in this nuclear era we cannot dare to stop short of this, if we want to live in the security of knowing that the exploitive economies, which put nations at one another's throats have been abandoned for good and that nations no longer act with the dangerous irrationality of adolescents in the schoolyard (as Norman Cousins once said), then we need to go after the causes of war which lie "in the minds of men."

But what he meant by "thought" is poetic shorthand for a whole way of thinking. If a true paradigm shift in science is a rare event which occurs only after a lapse of centuries, the shift we are speaking of is even rarer. It is a reorientation of the attitude of masses of human beings not only to a particular war, not only to war in general, but to our relationships with one another. It is a step forward not only in history but in biological evolution.

The questions that anyone concerned with peace today must ask, therefore, are two:

Are the times right for a perceptual revolution of this magnitude?

If so, what can I do to facilitate it?

War is no longer legitimate." If this is correct, we have reached what Kuhn would call an incipient paradigm breakdown which has not yet been carried through, for want of a paradigm shift.

And of course it is correct. Almost daily more disturbing anomalies confront those who still believe that political power grows out of the barrel of a gun—or a nuclear silo: The militarily strongest nation in the world is unable to control events in a tiny Southeast Asian country; to react effectively when its diplomatic staff is taken hostage in the Middle East; to preserve its economy; to maintain a foreign policy consensus at home, or the ability of its citizens to walk down the street in safety. The number two military power is in a similar predicament. Both add to their own and one another's insecurity with each new generation of wea-pons they produce to gain security.

To the majority, who are good Ptolemaeans, these problems, all of which are directly or indirectly produced by our commitment to military power, are still what Kuhn calls "puzzles" in a system which remains generally sound. But to a growing number of more thoughtful "Copernicans," from every walk of life, these prob-lems are not puzzles but genuine "counter-instances" showing that the entire paradigm in which they are sustained is wrong.

Can we get people to abandon the old paradigm? Only if and when we can get them to see a new one. As Kuhn observes, "once it has achieved the status of a paradigm, a scientist theory is declared invalid only if an alternative candidate is available to take its place.... The decision to reject one paradigm is always simul-taneously the decision to accept another." There can be no ques-tion of perception, evaluation and decision without a paradigm. That may be possible for the mystic, but it cannot be the ordinary process of human decision-making in science, and still less in politics.

Significant numbers of people will never stop making the decisions that lead inevitably to war, even if the resulting wars destroy everything they live for, until and unless they come to trust, understand and learn the use of an entirely new system of decisions that leads to peace. Fortunately, such a system is already known. These words of Albert Szent-Gyeorgyi point to it very clearly:

Any successful use of non-violence—the liberation of India from British rule is a conspicuous example—provides not only an arresting counter-instance to the old paradigm of injurious force but a clear indication of the new paradigm. It is based on an entirely different set of assumptions and a different system of human relationships. The term "nonviolence," like that of "para-digm," has been weakened in the social sciences and in its rare appearances in common parlance. But when truly understood it provides an entirely new conceptual system, and when correctly applied it provides mankind, in Gandhi's words, with "the grea-test force he has ever been endowed with."

Life theories, of course, are even harder to change than scien-tific theories. But the method history shows us is the same: First a few daring geniuses discern the new paradigm—a Copernicus, a Galileo, an Einstein, or in our case a Gandhi. Then certain opinion leaders take it up and demonstrate its power. In course of time it becomes the established frame of reference. That is why Einstein reckoned that if only 5 percent of the people would work actively on peace it would be achieved. Peace is not only inherently more

desirable but practically more workable than war, and as with any
such change, when opinion leaders begin to use the new paradigm what was at first the "lunatic fringe" can become insensibly, but rapidly enough, the carpet on which the majority takes its stand.

Our second question—what can I do—has therefore almost answered itself. Whatever we can do to increase the visibility of this new paradigm would be far and away the most effective contribution we can make to the establishment of a lasting peace. There are many ways to approach this adventure, but it seems to me that for all of us who are members of the intellectual commun-ity these ways would entail our learning the history of nonvio-lence, understanding the theory behind it, and—most importa-ntly—learning to practice it. What the Report from Iron Mountain (by Leoanard C. Lewin, Dell) had to say in 1967, that "up to now, no one had taken more than a timid glance over the brink of peace, "is still far too true.

At present, as we know to our cost, certain "buzz words" like "strength, preparedness "and above all "security", because they are thin disguises for military strength, and military security, play a major role in systematically misleading the policy decisions of people from the lowest to the highest political echelon. The most effective way to attack that kind of outmoded thinking is to point out, and where possible demonstrate, that there is a greater strength in cooperation and "mutual aid" than in belligerency; that any security worthy of the name can only come from not having enemies, not from threatening those we perceive as enemies. We have to know about the successful uses of non-threatening protective mechanism like nonviolent civil defense, and about other aspects of the nonviolent armamentarium to carry this point.

It is not uncommon to see a newspaper headline like "What U.S. Could Do to Iran" (meaning of course, what harm we could do to Iran). These questions are in the genre, "when did you last beat your wife?" They preclude any consideration of the one question which is in fact most important: Should we harm them? "The most effective way to open people's eyes to that neglected question is not to point out how we provoked the Iranians by harming them in the first place. (It may be true but it does not seem to be effective.) Rather, it is to demonstrate how cooperative and conciliatory behavior would be a more efficient way to deal with them than combative aggression.

Without a definite shift in our educational perspective we can hardly hope to effect a permanent shift in our world view. The things that unite the various people of the globe would have to be made more interesting than the present intellectual fascination with differences. Such a shift would prepare the ground for this opening of eye. We would have to know not only how the principle of peace is being violated all over the world, but precisely what it would take to make peace more stable, which as far as I can see can only be thoroughgoing nonviolence.

At the basis of the old, increasingly vulnerable but still very dangerous paradigm of force and violence there is the central assumption that people are separate, pretty much as they appear to the senses. We will never securely abolish war without chal-lenging this basic assumption; and the new paradigm does chal-lenge it. Nonviolence is based on the hypothesis that all life is one.

Just as Einstein opened the modern era in physics by challeng-ing the accepted axiom that time and space are absolute coordi-nates, so he and others now challenge us with the even more portentous hypothesis that men and women are not separate from one another and the rest of the environment. The day may be dawning—we should see to it that it does dawn—when words of Einstein on compassion are as much used as his famous formula expressing the inter-convertibility of matter and energy:

When Gandhi called nonviolence a science, as he often did, and referred to his own life as a series of experiments with truth he was not being metaphorical. Nonviolence is certainly science. It can be learned, and taught. It has a central hypothesis, as we have seen, from which a system of theoretical and practical constructs about the nature of reality have begun to be developed and tested. Its laboratory is the whole of life, and the scientists competent to explore it are, literally, all of us.

Much of the land that was contaminated by fallout from Chernobyl doesn't appear to have high levels of radiation. A Geiger counter registers levels only slightly about normal background radiation. This is deceptive. The radiation hasn't gone away; it's just been absorbed by the plants or the soil. People living in these areas would probably suffer very few radiation related problems if they could import their food from clean areas but for many of them that just isn't an option.

The idea that one would grow food in radioactive soil seems, at first glance, to be rather ludicrous. With all the arable land in the world why would anyone put seeds in the small amount of land that contains measurable amounts of radioactive isotopes? And who would be loony enough to eat the stuff that grows there?

It's difficult for Americans to understand the relationship that people in parts of Ukraine and Belarus have to the land they live on. Our culture worships upward mobility and even just plain old mobility. But that's not true for people like Danilo Vezhichanin, the mayor of Yelna, a village in an area north of the Chernobyl reactor known as the Polissa region.

Vezhichanin has lived in the village his entire life. His father, grandfather and great-grandfather all lived within a few kilometers. He says that he likes to sweat and work with his hands and that all he needs for a good life is rich soil, good water, a few pigs and a cow.

Yelna is a cluster of houses surrounded by small fields and lots of forest. There are bogs nearby that are considered bottomless because once you're in them, you just sink and sink and sink. Physically it's a few hours drive from Kiev but culturally, it's centuries away. An aid worker in Kiev described the drive into the Polissa region as "a Ukrainian Safari."

People in Yelna have TV sets and refrigerators but take away those few things and the place looks the way it has for ages. Water comes from open wells with crank up buckets. It's the sort of well that's always depicted as the destination for Jack and Jill. People get around using horse drawn carts and it seems that someone is always tugging a cow with a rope around its neck from one place to another.

The deep green forests and labyrinths of swamps have been used over the centuries to conceal rebels of all stripes. Conquerors would find empty villages at tax time because the inhabitants would disappear into the marshes without a trace. During World War II, or the Great Patriotic War as it is called in this part of the world, anti-Nazi partisans found food and shelter with the help of the locals.

The Chernobyl plant was cited in the Polissa area because it provided a good supply of water. People in Yelna didn't know much about the plant and didn't think about it at all. It wasn't until six months after the accident that the people in Yelna realized that their village had been contaminated by Chernobyl. They'd seen the reports on TV but thought that it all happened too far away to affect them. People in the village began to get sick. They got headaches and joint pain but nothing that seemed serious.

The Soviets gave them the option to leave. They offered to buy up their houses and resettle them in other areas. But the people in these villages didn't have a lot of either faith or trust in the Soviet government. There were still people who remembered the famines caused by Stalin's collectivization.

Also, during the Stalin years, whole families had been declared "enemies of the people" and shipped off to Siberia. When they got the chance they returned back to their home villages. Those stories were still vivid in the hearts and minds of the people living there. They were understandably suspicious of the government's offer to relocate them.

About 500 families left the region but nearly everyone in Yelna stayed. Vezhichanin considered moving but he had no idea where he would go. He'd always lived in Yelna. He had a warm house and some land to work. He'd visited Kiev and couldn't stand the place. Like other people in the village, he stayed and hoped the problems would just go away.

While the Soviet Union was still intact, the local grocery store had shelves full of clean food from other regions and the residents were given a small stipend as compensation. It wasn't much but it was enough to buy uncontaminated food.

When Ukraine got its independence in 1991, all of that went away. The Ukrainian government can't afford to buy houses and there is no private market for the real estate. No one wants to buy radioactive land.

They still get a stipend of 2 hryvnia 10 kopecks every month for every person. It amounts to a little over 40 cents. They call it coffin money.

There is no industry. The only jobs in the area were in the collective farm. Before the accident the collective sold its produce and generated income for the area. Now there is a ban on agricultural products from Yelna.

Since they can't sell their products legally, the only source of income is a thriving black market in forest products. During the summer, mushrooms, blueberries, blackberries, currants and cranberries grow in the forests and bogs. Women collect the bounty in handmade baskets. They wade up to their necks in the swamps to harvest cranberries. On average, they can make 600 hryvnia (a little over US$100) in the summer months. Ambitious workers can make up to 800 hryvnia.

On summer evenings, trucks pull into the village and the drivers buy everything the women have harvested. Everyone knows the produce is full of cesium. When asked about this Vezhichanin shrugs and says that the people who buy it have equipment to test the radiation levels. In the next breath he'll say that the trucks take the mushrooms and berries into countries like Poland where no one is checking radiation levels.

It isn't just people living near Chernobyl who eat radioactive food. The mushrooms are dried and packaged. The berries are made into preserves. The resulting products are shipped all over Europe and possibly imported into the U.S. After all, who would think of checking the radiation levels in food?

They know radioactive food is a problem in Moscow. Food inspection includes testing for radiation. In the first half of 2005, 356 kilograms (785 pounds) of radioactive berries and mushrooms were confiscated from Moscow markets.

CHAPTER 7: Film and Television Productions

"Films deal with the emotions and reflect the fragmentation of experience"

— Stanley Kubrick

A French organization asked her to distribute a shipment of humanitarian aid to Rovne in Ukraine because they knew she was honest and the aid would get to the people who needed it. She left her children with her husband and went off to supervise. While she was gone a fire started in the corridor outside their apartment. They threw whatever they could out the windows only to have it stolen by looters. No one was injured but when she got back she got a phone call telling her that next time her ashes would be spread all over.

Now it was out and out war. Svetlana and her family moved in with friends. They had to apply to Zaretsky for housing. He told her that she should get a house in Israel. Between that and the fire Svetlana was ready to give up. She called a friend in Gomel and cried on her shoulder about everything. Her friend worked in the TV station and promised that if Svetlana held a meeting in the central square of Gomel a camera crew would show up.

Svetlana put together all the documentation she had on Zaretsky. Because of all the evacuations, it was difficult to get housing in Belarus at that time.

The government doled out apartments based on a person's passport. Zaretsky had 5 fake passports that he used to get apartments in Minsk, Gomel and other areas. Svetlana had the addresses of the apartments and the passport numbers.

When she arrived at the square the militia was already waiting for her but so was the TV news crew. While she was being dragged away in handcuffs, she screamed out the passport numbers and addresses. The scene was shown on national television. News programs in Moscow and St. Petersburg picked it up. Zaretsky had a lot to answer for.

Three days after the TV broadcast a teacher called Svetlana with another warning. The school administrators had asked the teachers to gather information about Svetlana's children. They wanted to know what they wore to school, what they ate at home, who took care of them when

Svetlana was away. They wanted to terminate her parental rights.

Svetlana phoned the head of the Children's Fund, a state organization in Minsk. When she reeled off the whole story he told her to calm down and he'd handle the problem. The next day he phoned her and told her to turn on the TV. There he was demanding to know why, "they keep booting this lady."

The school administrator called and apologized and she swore that she had no idea why the government wanted the information. Svetlana and her family got their housing but it's still in a contaminated area. Her daughter has a serious thyroid problem and her husband died in 1994.

After Belarus gained its independence Svetlana hoped for a change in the local government but Zaretsky was left in charge. She sent letters to President Lukashenka hoping that he would look into the matter. Lukashenka looked over the information but declined to act when he found out that the National Party, his political opponents, backed Svetlana.

Eventually Zaretsky was transferred to another area and replaced with an honest administrator. Zaretsky died in 2004. Svetlana is still distributing aid for various organizations even though she's less active than she used to be. The old people in the abandoned villages say that the only person who brings them something is Svetlana. No one else helps them. And for Svetlana, that's all the reward she needs, which is a good thing because it's all she's going to get.

Svetlana has eight children and when I interviewed her she was in her 50's. She hadn't worked in 5 years because she couldn't find a job. She would soon be eligible for retirement, which means she'll get a government pension but it will be very low because it's tied to her earnings. Like everyone else in the area, she had a large garden and keeps chickens.

She has a modest house with a fenced yard that's taken up almost entirely with her vegetables and flowers. With the sun shining, the chickens clucking around in the yard it looks idyllic, as long as you don't think about the radiation.

She insisted on feeding us and it was an amazing spread. There were several kinds of salads, boiled eggs, mashed potatoes, smoked salmon, chicken, fried zucchini, ham, wine and, of course, vodka. The crew I was traveling with that day consisted of two photographers, a video cameraman, a translator, an employee of Belrad and our driver. She made sure that every one of us ate until it hurt. Her generosity and hospitality is legendary among the aid workers who've visited her.

After dinner Svetlana brought us to a nearby village that was officially evacuated. A woman wearing a headscarf and a worn housedress came out to greet us but refused to let me interview her saying, "I'm old. I don't know anything." She brought out some apples and insisted that we eat them. Even though we'd just finished a huge meal, it would be rude to refuse. While we ate the radioactive apples, people wandered in from nearby houses.

First an old man sat down and stared at us as if waiting for a show to begin. A couple of middle-aged women talking conspiratorially to each other walked over. Then three children came over, giggling and whispering to each other. I was astonished to see the children. It's usually older people, like Grigory, who move back into areas of high contamination. Their reasoning is that since they're old, they'll die of something else before the radiation can get them since most of the cancers associated with exposure have latency periods of 5 to 20 years.

More people came over and finally a middle-aged man introduced himself as the representative for the village. We could interview him.

He had been born in the village but he'd lived in Gomel for many years and worked in the militia. Now he's retired and living on a pension so he moved back the village.

I asked him about the children and an older man from the audience piped up. They were his grandchildren. He said that the only reason he and his wife had moved back to the village was to help out their children. "They live in Gomel," he said," but they can barely get by. I plant potatoes for them and I have a cow so I can give them milk."

Apple trees grow in every yard. They sell the fruit to a local distillery that makes cheap wine. When I asked if they were worried about radiation in the food, they all laughed. The old man said, "If the apples are radioactive then the alcoholics will die sooner." He wasn't worried about his grandchildren eating the food because he knew that the food in Gomel was just as contaminated. Most of the produce sold in the Gomel market was grown locally.

I'm not sure why the villagers felt they needed a spokesman because they couldn't resist speaking up. I'd ask him a question and then the old man would call out an answer, his wife would add her two cents, the middle-aged women would add their opinion, the children giggled, the chickens squawked, the dog barked and a good time was had by all.

When I asked them what they knew about the plant before the accident, the spokesman said, "We didn't know anything about atoms." They were vaguely aware that some sort of power plant was nearby but they knew that there were plants like that all over the Soviet Union so it had to be safe.

After politely eating a few more radioactive apples and pears, we drove to an area that was so hot it had a checkpoint with a guard. Svetlana said that if we wanted to go into that area we could easily avoid the checkpoint by using another road. The locals knew this and often people went into the hot area to pick berries and mushrooms. We declined the invitation.

As we were driving back to Svetlana's house she pointed out a field where a combine was harvesting grain. She told us that it's illegal to grow anything there because the area is too contaminated. She cautioned us not to get out of the car and take pictures because that might result in a visit from the local authorities, who were getting kickbacks for allowing people to grow food there.

We had been driving around with Svetlana's 14-year-old son and he asked me the most difficult question so far. With a huge smile and the enthusiasm any kid would have for a local attraction he asked, "So, how do you like Chernobyl?"

He'd lived with the radiation his entire life and accepted it as part of normal existence. He saw it as something that made his village special and probably had no idea what it would mean to him in the long run.

A normal part of growing up is learning that there is no Easter Bunny, Santa Claus or tooth fairy. But children who grow up in the zone have to come to terms with some very ugly realities early on in life. We visited a school that had equipment for measuring the radiation in food. Children as young as six were taught about the radioactive elements in the environment and were encouraged to bring in food from their home gardens for testing.

The principle gave us a tour of the school. She was a wiry, energetic woman in her fifties with permed hair and dirt under her fingernails. She wore a straw hat with a bow, jeans and a T-shirt. She kept everyone in line. She showed our driver where to park, reeled off an extensive list of people she wanted us to interview and interrupted our translator every time he opened his mouth, all at the same time.

Since so few children had been born in the past decade, most classrooms were empty. One had been turned into a local museum. Along with the wooden farm implements and traditional dresses, there was an entire wall with gruesome pictures devoted to the story of a local woman who fought against the Germans during World War II. She was captured, tied to a horse, dragged through the streets to the cemetery and then shot.

Another exhibit had an array of helmets, guns and land mines plucked from the surrounding fields after World War II. But nothing beat the formaldehyde soaked three-legged chicken in a jar. It was born after the Chernobyl accident.

From the first day of school, children learn that their hometown was a bloody battlefield and everything that grows here is probably full of poisons.

We spoke with a young man who taught at the school and planned to stay and raise a family. He'd grown up in the area, left for college but came back to teach. He doesn't feel he has any choice but to stay. Teaching jobs are just too hard to find in Belarus.

He was five years old at the time of the accident. He said that his parents tried to make sure he ate only clean food but they had to drink water from the tap.

He's had one operation for thyroid cancer but the latest x-rays showed that he now has two more tumors. He's taking medication but if it doesn't work he'll have to have another operation.

The principle said over the past few years it seemed that fewer children were getting sick. She was sure that it was because they went to clean areas for two months every summer. When they got back from their vacations, they usually had very little radiation left in their bodies.

The long-term outlook for the village wasn't good. She estimated that every year about 12 people die and about 5 babies are born. Four of the five newborns have to spend a month in a special center in Gomel because their immune systems are compromised. But she wasn't sure if the problems were related to Chernobyl or the amount of alcohol consumed by their mothers.

It was dark outside by the time we left the school. As we drove past the fields we could see several fires burning. They burn the fields after the rye harvest, even though the fires give off radioactive smoke. It's what they've always done.

The smoke persisted even after we got back to Gomel and I asked our driver, Slava, why there was so much smoke. He thought it was the peat bogs. Swamps surround Gomel and when the bogs dry out in the summer, the peat moss spontaneously combusts and underground veins of peat burn uncontrollably.

Slava then casually mentioned that a short while after the accident, when he was a college student, he had to help with the potato harvest near the exclusion zone. This was a standard practice in the Soviet Union. College students were required to help the farmers bring in the harvest every year.

Slava insisted that it was fun because they were fed lots of milk and bread. It was like having a picnic for a couple of weeks. Aside from the radiation, there was only one other little problem. The bogs were burning.

The fires smoldered underground, tunneling under the potato fields. Every once in a while a truck or a student would disappear when the ground gave out underneath them. They would fall into a pit of burning peat moss. No one survived such a fall because they were baked alive.

Slava has survived digging up radioactive food, inhaling radioactive smoke, burning bog pits and eating contaminated food for 18 years. This is a very tough place to live.

Conclusion

*"... for He knows how we are formed,
He remembers that we are dust"*
—Psalm 103:14

Fifteen years before the worst nuclear accident ever, teams of spirited pioneers were breaking through the frosty marshlands in Pripyat. The Chernobyl plant was an iconic manifestation of the age of science and technology emerging in the

Mother Land. It was to be the pinnacle of the Soviet Union, praised for ingenuity and scale, with immense RBMK reactors feeding into the grid for the good of the people.

It's an ironic twist that the celebrated nuclear city birthed from nothing would implode to less than nothing at the peak of its success. Subsequent to illusions of grandeur and impossible dreams, its architects were forced to dismantle the pinnacle of their life's work, exquisitely adorned but now void of hope. Chernobyl became radioactive, literally and metaphorically, ending lives, and killing dreams.

Human ego fueled the greatest nuclear disaster in history, adorned by falsities, bureaucratic labyrinths, and a collective blind eye to the dangerous truth of their golden calf, nuclear energy. It was a culture of plausible deniability, run by commitment-phobic leaders, dispensable pawns of the system themselves.

The designers of the Soviet nuclear reactors were far more concerned with keeping up appearances (and miraculously lowering costs) than with exploring the safety parameters of hypothetical emergencies. Their reactor was a success, their political positions secured, and their status as national heroes already penned into the parchment.

A clear dichotomy is captured in the dry tone of the revealing narrative. The heart wants to celebrate the ingenuity of the apparent underdog—the poorer, oppressed scientific community in the USSR—in their goal to compete against other superpowers like the United States. Sputnik, for example, was the boost they needed to establish credibility with the masses, assuring the Soviet citizens of the nation's competitive edge.

It was all a lie, though, a political house of cards with expendable pawns in an earth-shattering game to see who survived. The 'red tape' of the

USSR blended into the scarlet May Day parades even as the air roiled with radioactive particles settling on the heads of the innocent. The cognizance of how political power has unwittingly threatened the very existence of humanity in a nuclear arms race is sickening.

Blame isn't reserved for dictatorships, communist states, or irresponsible monarchies, either. The free world, too, has a seat at the table of doom. The tension between communist ideals and ambition is palpable as the story builds, pulling the reader right into the fray. How any nation could deteriorate to that point of reckless abstraction is unfathomable, yet the world is still fixed upon nuclear energy as the only hope in the battle against climate change thirty years after Chernobyl.

Fear and pride ruled the hierarchal society with misguided faith in the greatness of the motherland. Blind obedience to superiors coupled with severe consequences for perceived dissenters echoes more recent tragedies, like the controversial South Korean plane crash in San Francisco in 2013. The only difference is that in Chernobyl the stakes of failure were much higher and involved an entire system of people unwilling to do what was right.

One envisions Higginbotham as the proverbial Perevozchenko, the reactor supervisor, as he urges the reader, his Yuvchenko, not to stare into the nuclear hell that teetered on the precipices of the Chernobyl disaster. It may just save us a life bewildered by the depths to which humanity sunk in the quest for power.

The book reads like the captivating investigative documentary style one expects of such an infamous catastrophe, slow at times but building in intensity. The fascination of the incredulous author is splashed liberally between the power and tragedy, and he manages to eat the radioactive elephant's foot one engaging bite at a time.

The bizarre truth, carefully laid out from unsealed government reports, witness statements, and public record, pulls at the flailing Soviet system, picking it apart, splitting atoms and hairs interchangeably.

Real people (most of whom are still alive) are poignantly central to the account, the human collateral of the Iron Curtain cinching a phenomenal story over just another mediocre fact-finding mission.

Bibliography

- Alexievich, S. (2006) *Voices from Chernobyl: The Oral History of a Nuclear Disaster. Trans. Gessen, K. London: Picador.*

- Charles River Editors. (2014) *The Chernobyl Disaster: The History and Legacy of the World's Worst Nuclear Meltdown. CreateSpace Independent Publishing Platform*

- Deshusses, H.P. (1997) *La radioactivité dans tous ses états. Geneva: Georg Éditeur.*

- Medvedev, G. (1990) *La vérité sur Tchernobyl. Paris: Albin Michel.*

- Rubeau, D. (2001) *Catastrophes et accidents nucléaires dans l'ex-Union soviétique. Nanterre: EDP Sciences.*

- *International Safety Advisory Group. The Chernobyl Accident: Updating of INSAG-1: INSAG-7. Vienna: International Atomic Energy Agency, 1992*

www.ingramcontent.com/pod-product-compliance
Lightning Source LLC
Chambersburg PA
CBHW071831080526
44589CB00012B/982